FROM SPARK TO SPACE
A Pictorial Journey Through 75 Years of Amateur Radio

Edited by Debra A. Jahnke and Katherine A. Fay, N1GZO

Graphic design and layout by Sue Fagan

Typesetting by Michelle Chrisjohn, WB1ENT

Cover graphic by Bob Rich
(The Bob Rich Airbrushing Studio)

Contributors
Sheldon H. Ball, KC1MP
Charles Bender, W1WPR
Jon Bloom, KE3Z
E. Laird Campbell, W1CUT
Bart J. Jahnke, KB9NM
Kirk Kleinschmidt, NT0Z
John Nelson, W1GNC
Paul L. Rinaldo, W4RI
David Sumner, K1ZZ
Perry F. Williams, W1UED
Mark J. Wilson, AA2Z

Published by
The American Radio Relay League
Newington, CT USA 06111

Copyright © 1989 by

The American Radio Relay League, Inc

Copyright secured under the Pan-American Convention

This work is publication No. 114 of the Radio Amateur's Library, published by the League. All rights reserved. No part of this work may be reproduced in any form except by written permission of the publisher. All rights of translation are reserved.

Printed in USA

Quedan reservados todos los derechos

ISBN: 0-87259-265-0

$20.00 in USA and Possessions

First Edition
First Printing

CONTENTS

- 5 — A Message from the President
- 7 — Foreword
- 8 — Our Beginnings
- 9 — Hiram Percy Maxim
- 15 — W1AW—Amateur Radio's Dream Station
- 24 — Cartoons Through the Years
- 29 — ARRL...The Dream—The Reality
- 35 — The Early Years
- 43 — The Thirties
- 51 — The Forties
- 59 — The Fifties
- 67 — The Sixties
- 75 — The Seventies
- 83 — The Eighties
- 91 — Field Day Through the Years
- 93 — The Future of Amateur Radio

A Message from the President

"From Spark to Space" truly describes the historical path of Amateur Radio and the American Radio Relay League. Two men, Hiram Percy Maxim and Clarence Tuska, carefully laid the groundwork for the traditions that have guided amateurs for the past seventy five years: public service, technical progress, and operating skills.

Through the twentieth century, amateurs have been innovators and experimenters who overcame adversity and obstacles. Challenges were not considered stumbling blocks; rather, they were accepted with great relish.

While our ideals have remained the same, our methods certainly have progressed and changed. Technological advancements have allowed us to widen our scope of communications worldwide and beyond. "State-of-the-art" quickly becomes commonplace, and advancements become more and more extraordinary. An amazing example was the recent NASA *Voyager* mission, where slow-scan television enthusiasts were privileged to view almost immediate retransmissions of the planet Neptune carried on an amateur band.

Anniversaries are a time for reflection and reminiscence. This is a very special year for the League. We have been serving the Amateur community for 75 years and we are proud of the role the League has played in the advancement of Amateur Radio.

Anniversaries are also a time to look forward. If past achievements are any indication, our future possibilities are endless. The coming decade and the coming new century will be an exciting time to be a radio amateur.

But then, it always has been.

Larry E. Price, W4RA
President, ARRL

Foreword

The 75th diamond anniversary of the American Radio Relay League was marked in 1989. In July, the Hiram Percy Maxim memorial station, W1AW, was rededicated after undergoing extensive renovations. One cannot help but feel that "The Old Man" would be proud of "his" station and of Amateur Radio as well.

HPM was a man of vision. His mind was on the future. His biography, which appears on page 9 attests to this. "Rotten" (a favorite expression) is how he would refer to our setbacks. But he would have high praise for our achievements.

"From Spark to Space" displays pictorially some of the many changes Amateur Radio, and the League, have witnessed through the years. You can sit back, smell the ozone and hear the crackle of a spark gap generator. Turn a few pages and see "Oscar I" launched into space. Then go back and spend some time in your own favorite decade. Enjoy your journey.

73,

David Sumner

Newington, Connecticut
November 1989

David Sumner, K1ZZ
Executive Vice President and Secretary

OUR BEGINNINGS

It was December 12, 1901. In an old barracks on Signal Hill at the mouth of the harbor at St. John's, Newfoundland, two men sat before a strange apparatus. Over their ears they wore headphones. From their equipment a wire led to a kite floating four hundred feet overhead. They listened intently. Suddenly, at just about noon, they heard a faint, crackling series of sounds. It came again. The sounds were unmistakable—three quick buzzes—repeated over and over. The letter "S" in Morse code. It was a long-awaited signal emanating from Poldhu, Cornwall, on the southwest tip of England. For the first time, a signal had been transmitted and received across two thousand miles without wires. The man responsible was Guglielmo Marconi, whose name quickly became famous around the world.

Our next scene takes place in March 1914. The locale, Hartford, Connecticut. Another inventive genius—in a quite different field—and also an ardent radio amateur had a problem. His name was Hiram Percy Maxim and he was trying to contact a fellow amateur in Springfield, Massachusetts, less than thirty miles away, without success. On an impulse, he switched his effort and made contact with another amateur located midway between Hartford and Springfield, and arranged to have him relay the message. The next day, still thinking about the experience, it dawned on Maxim that here was exactly the spark that could launch an organization of radio amateurs across the country.

And so it began. The American Radio Relay League was born with Hiram Percy Maxim as its first president and a young man named Clarence Tuska as its secretary, and later as the first editor of *QST* magazine.

HIRAM PERCY MAXIM

Hiram Percy Maxim was the third member of the Maxim family to achieve fame through his inventive genius. His father, Sir Hiram Stevens Maxim, became a British subject and was knighted by Queen Victoria in 1901 for his invention of a Maxim Gun, the first practical machine gun that fired 600 shots per minute. His uncle, Hudson Maxim, introduced Maximite, a powerful explosive used in armor-piercing projectiles.

Hiram Percy Maxim began his notable career in mechanical engineering with his graduation from the School of Mechanical Arts, Massachusetts Institute of Technology, class of 1886. He was the youngest member in his class at seventeen.

He gained valuable experience during the next several years working as a draftsman, assistant engineer, chief engineer, and superintendent.

He excelled in many fields. The "Columbia Gasoline Carriage," winner of the first automobile race held in America, at Branford, Connecticut in 1899, was designed by Mr. Maxim. In later years he became avidly interested in aviation, astronomy, cinematography, photography, yachting, and of course, Amateur Radio.

In 1907, the Maxim Silencer Company was organized with HPM as president. In 1909 he introduced his invention, the Maxim Silencer, which made gunfire noiseless. This invention brought him worldwide fame and started him in his life work, the abatement of all noises detrimental to human life. A less publicized invention was the "Maxim Window Silencer," which allowed complete room ventilation but kept out the offensive noises that ordinarily come through open windows.

It is significant that, with his great talent for organization, he founded and until his death remained president of the preeminent organizations in two of his hobby activities— The American Radio Relay League, International Amateur Radio Union, and the Amateur Cinema League.

The greatness that was Hiram Percy Maxim's was well expressed by the *Hartford Times* on February 18th, 1936:

"He had a boundless enthusiasm for everything that was new. Unlike most scientists he was not content with a purely materialistic view of the universe. In recent lectures he had said that the more he familiarized himself with all that science had discovered, the greater his respect for the orderliness of it all and the stronger the conviction that behind the order must be some supreme force. Knowledge made him neither discontented nor pessimistic. Life remained for him to the end a great and exhilarating adventure. He was a remarkable man, a choice spirit."

HIRAM PERCY MAXIM

Born at Brooklyn, N.Y., September 2, 1869 son of Sir Hiram Maxim, inventor of automatic firearms.

Educated at Massachusetts Institute of Technology, Class of '86. No post-graduate work.

Business Connections:

Sun Electric Co., Woburn, Mass., 1886-1887, as draftsman.

Ft. Wayne Jenny Elec. Co., Ft. Wayne, Ind., 1887-1888, as draftsman.

W.S. Hill Elec. Co., Boston, Mass., 1888-1889, as draftsman.

Thomson Electric Welding Co., Lynn, Mass., 1889-1895, as draftsman and, later, assistant engineer.

Pope Mfg. Co., Hartford (and its successor the Electric Vehicle Co.), 1895-1901, as Chief Engineer.

Westinghouse Electric & Mfg. Co., E. Pittsburgh, Pa., 1901-1903, as Vehicle Motor Engineer.

Returned to Electric Vehicle Co., Hartford, 1903-1907, as Chief Engineer.

Organized Maxim Silencer Co., 1907, as President, and held this position until his death.

Organizations:

Founded the American Radio Relay League, Inc., in 1914 (an association of radio amateurs) and served as its President until his death.

Founded the Amateur Cinema League in 1926.

Instrumental in forming the International Amateur Radio Union, an organization of national amateur radio societies.

One of the principal organizers of Aero Club of Hartford, and its President from time of organization in 1907 until 1926.

Former Chairman Hartford branch American Society of Mechanical Engineers.

Former President Technology Club of Hartford.

Former member Executive Committee, M.I.T. Alumni.

Former President Aviation Commission of the City of Hartford.

First President, Engineer's Club of Hartford.

Former Director, Hartford Yacht Club.

General:

Honorary Degree, Doctor of Science, from Colgate University, 1924.

Inventor of Maxim Silencer.

Chosen field was sound, but was also keenly interested in radio, was a pioneer in the automobile industry (early part of 20th century), and was active in aviation and the amateur cinema movement.

A Letter to T.O.M.

Dear Mr. Maxim:

I suppose there are some who think it rather strange that I should be writing a letter to someone I never knew, and who has been a Silent Key for over 50 years. I would reply by requesting that they reserve judgment until after reading my letter.

At the time you died, I had just passed my 14th birthday. Please note, I did not say "celebrated." When you were from a poor family in the midst of the Depression, you had very little to celebrate. We youngsters of the '30s knew a lot about friendship, though. When you passed on, Mr. Maxim, I felt I had lost a friend and I was very, very sorry.

My early years were so very typical of most hams and SWLs of that period. I lived in a small town of about 500 people; the nearest ham was more than 30 miles away. I was building crystal sets and experimenting with radio and electrical circuits long before I knew anything at all about Amateur Radio.

My first exposure to *QST* came when I discovered a few copies in a box of other magazines a visiting relative brought us. I read them over and over, and still have them today.

I heard of a ham, W5FKT, who lived in Springdale, some 30 miles away. I talked my Dad into driving me to Springdale, so that I might actually get to see a ham station. Clyde, W5FKT, opened the door and invited me in. I will never forget the beauty of the 250-watt rack and panel transmitter and the Skyrider sitting on the table. Clyde turned the receiver on and I was amazed at the volume and clarity of the signals. Turning on the transmitter he called "CQ 160"; the 866s furnishing power to the class B modulator flashed with their beautiful blue glow as he spoke into the mike. He was answered by a ham over 50 miles away, and when he handed me the mike I was speechless. I managed a few mumbled words and left walking on clouds. I also left with a armload of *QST* magazines and a license manual, loaned to me by W5FKT.

Thus it was that I read of your death, Mr. Maxim. I also learned many other things, like good operating practice, and the fate awaiting those who strayed too far from the straight and narrow—the dreaded Wouff Hong and Rettysnitch.

During those early years I read and reread every old *QST* I could lay my hands on. My favorite articles were the ones by "The Old Man." I also read the construction articles until the schematic diagrams were committed to memory. When I should have been reading the classics for English Literature classes, I was drawing circuits for my first transmitter.

When I called my first CQ, I felt you were standing behind me, Mr. Maxim, just waiting for me to call CQ more than four times without signing so that you could take the Wouff Hong to me. When I heard W5HHR answer my CQ, I was so excited I could not finish the QSO. I made a real mess of it, but was elated that I was finally a ham.

I could not muster up enough courage to try another CQ that day, but I set my alarm for 3 AM and tried again the following morning. Before going to sleep, I spent two hours reading my well-worn old copies of *QST* again. Your words assured me, and gave me hope and encouragement. By morning I had worked two more stations, actually completing both QSOs. I then went outside and sat on the front porch watching the sunrise, thinking how wonderful ham radio was and how lucky I was to be ham. I promised myself I would learn to send good code, that I would constantly try to improve my operating habits, and would at all costs try to avoid the wrath of The Old Man.

Now, 50 years later, I am just about the age you were when you became a Silent Key. You are still with me every time I turn the rig on. Only last week, while tuning the Extra section of the 20-meter band, I heard a station call "CQ" 26 times before signing his call, and I thought of you and the Wouff Hong.

Please don't get the idea that I think all you do is look down from that great shack in the sky and frown. Often, I see your face when it is nothing but smiles; while gently puffing on your pipe, you are nodding your head, showing how pleased you are. When do I see this look? When an old-timer takes the time to help a newcomer get his ticket. When some hams take time to help others put up a new antenna. When rare or semirare DX stations move into the Novice and General bands to give them a chance at some new DX. When the CW op who normally runs at 30 WPM+ slows up and ragchews for a half hour with someone who can't get his speed up over 15 WPM.

I don't claim to know the answer to our problems, Mr. Maxim, but somehow I feel that every generation of hams need an Old Man, and maybe even a Wouff Hong. Someone to look over our shoulder and constantly remind us to do our best, to improve our fist, to observe good operating habits and to always be considerate of our fellow hams.

So you see, Mr. Maxim, I feel I do know you and that you've been keeping an eye on us all these years. I just wanted to say thanks for doing so much during your lifetime that, 50 years after your death, our hobby is still healthy and strong. It has been a rewarding experience to have known you.

73,

Bruce Vaughan

Bruce Vaughan, NR5Q

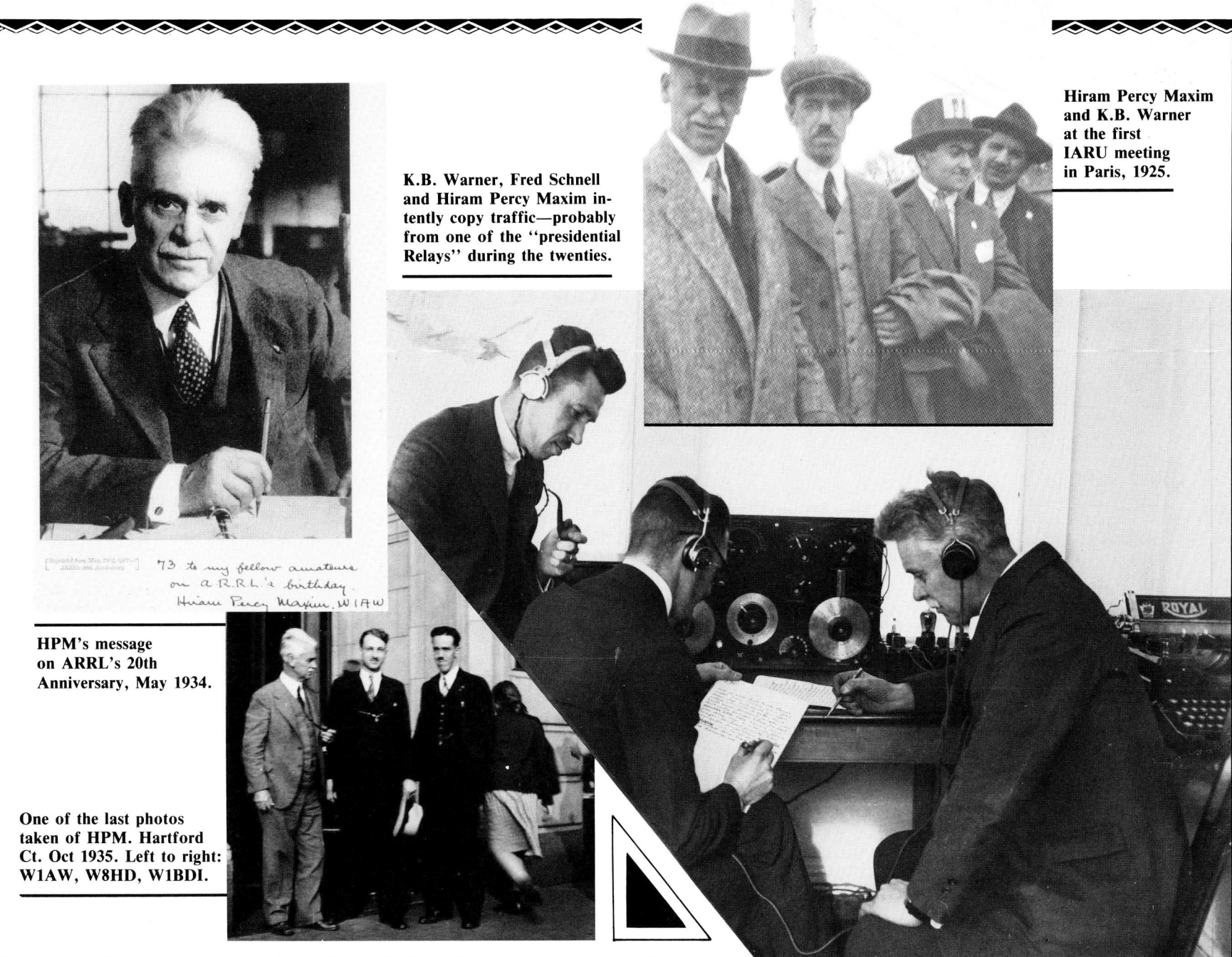

K.B. Warner, Fred Schnell and Hiram Percy Maxim intently copy traffic—probably from one of the "presidential Relays" during the twenties.

Hiram Percy Maxim and K.B. Warner at the first IARU meeting in Paris, 1925.

HPM's message on ARRL's 20th Anniversary, May 1934.

One of the last photos taken of HPM. Hartford Ct. Oct 1935. Left to right: W1AW, W8HD, W1BDI.

TUESDAY, FEBRUARY 18, 1936

Hiram Maxim Dies; Invented Arms Silencer

Taken From Train in Colorado With Throat Infection, He Succumbs at 66

Was Last of Triumvirate

Experiments on Car Muffler Led to Celebrated Device

By The Associated Press

LA JUNTA, Col., Feb. 17.—Hiram Percy Maxim, inventor of the Maxim silencer for firearms and numerous ordnance devices used in modern warfare, died in Mennonite Hospital here today of a throat infection. He was sixty-six years old.

Mr. Maxim was en route by train to the West Coast with his wife last week when he became ill. He was taken from the train and remained in the hospital here ever since.

With him were his wife, Mrs. Josephine Hamilton Maxim, and their children, Hiram Hamilton Maxim, of Hartford, Conn., and Mrs. John T. Lee, of Farmington, Conn. Also surviving is a sister, Mrs. George A. Cutter, of Dedham, Mass., and four grandchildren. The Maxim home is at Hartford.

Funeral services will be held on Friday morning in Hagerstown, Md., Mrs. Maxim's native city.

Last of Inventor Triumvirate

Hiram Percy Maxim was the last surviors of a notable triumvirate of inventors given to the world by a New England family. He was best known for his Mixim silencer just as his father, Sir Hiram Maxim, was best known for the Maxim machine gun, and his uncle, Hudson Maxim for Maximite and other high explosives.

The family was of French Huguenot

Inventor and Engineer

Blackstone Studios
Hiram Percy Maxim

extraction. Religious persecution forced its members to emigrate to Kent, England, and about 1650 they moved on to the British colonies in America, settling in Plymouth County, Mass. The first of the family in America was Samuel Maxim, great-great-great grandfather of Hiram Percy. The Maxims moved to Maine, and then moved again to Brooklyn, where Hiram Percy was born.

By that time, September 2, 1869, the family already was justly famed. Various members had served with gallantry in the French and Indian wars, the American Revolution and subsequent conflicts. Hiram Percy, his father, and his uncle built brilliant careers.

They had ideas and notions of how to do things with metal and wood fabrics. Their most successful inventions dealt with firearms and explosives. Their efforts to make things for peaceful humanity were not as richly rewarded.

Brothers Estranged

Hiram S. and Hudson were estranged and bitter brothers during the last years of their lives. Hiram by then had become a naturalized and knighted Englishman, feeling, perhaps, that he was not sufficiently appreciated in his own country. As Sir Hiram, he charged that Hudson had pirated his formula for smokeless powder. Some of his friends later added the charge that Hudson was posing in the United States as the inventor of the Maxim machine gun.

Hudson answered this with a biting public statement in which he assured the world that he had never invented a gun and never intended to.

He couldn't have hit a sorer spot, for Sir Hiram had become a decided pacifist, given to consoling himself with the fallacious idea that his inventions of guns and cannon had made war so horrible that never again would any nation dare to take up arms. He died at London in the fall of 1916, after what must have been a very puzzling last two years. His products had been serving the Central Powers and the Allies with equal effect and reliability.

Hudson Author of Atomic Theory

Hudson Maxim outlived his brother by eleven years, and when he died he was praised for his pioneer theory of the complex nature of the atom, and for his early process of color printing for newspapers, as well as for Maximite, the explosive which did not go off until the shell had pierced the enemy battleship's steel side. Much of his work was done in the laboratories of the E. I. du Pont de Nemours Company, of Wilmington, Del. But he found time to write a book, taking for his subject the modest field of "The Science of Poetry and the Philosophy of Language."

Hiram Percy had a tradition to maintain, and no one could deny that he made a good job of it. He attended the Brooklyn schools and made a splendid record at Massachusetts Institute of Technology, where he was the youngest member of the graduating class of 1886.

He had specialized in electrical engineering, and found work immediately with the Jenney Electric Company, of Fort Wayne, Ind. A year later he returned east and took a position with the W. S. Hill Company, of Boston. He held positions of successively greater importance with the Thomson Electric Welding Company, of Lynn, Mass., the American Projectile Company, of Lynn, and the Electric Vehicle Company, of Hartford, Conn. He was the founder and president of the Maxim Silencer Company, of Hartford.

Turned Out Electric Automobile

His promotions almost invariably were the result of his inventions. He developed methods of torpedo propulsion for the American Projectile Company, and while doing so became interested in the propulsion of road vehicles. Around 1894 he turned out a three-cylinder, air-cooled gasoline tricycle, and two years later introduced the Columbia electric motor carriage, the first practical electric automobile. It had a speed of twelve miles an hour and ran thirty miles on one charge.

Mr. Maxim designed a complete line of electric automobile control and charging devices for the Westinghouse Company, of Pittsburgh, and for a time he tried to get a foothold in the gasoline-engine automobile field. That venture failed, but led indirectly to his most celebrated invention.

Muffler Led to Silencer

He had to develop an experimental muffler to silence the explosions of the gas engine. He then tried the same thing out of a gun.

The problem was very different, however, for the successive rings which could dampen and silence noise in the exhaust pipe of a car were unsuitable for a gun barrel. A solid projectile, the bullet, had to pass through the barrel at high speed and without the slightest obstruction, but the gases and noise behind it had to be trapped and hushed.

Mr. Maxim found the answer in a round tube, to be attached to the muzzle of the gun. Within the tube were twelve small rings, each having a hole in the exact center slightly larger than the bullet. The inner edge of each ring was bent outward and down into a curve. Then, as the bullet passed through, the gases driving it and the sound waves created by the explosion of the charge were caught by the rings and given a whirling motion. A similar vortex was created by each successive ring, and the inner walls of the tube, against which the sound was thrown successively damped the sound until at the outer end, after the twelfth ring had been passed, from 90 to 97 per cent of the noise had been eliminated.

When Mr. Maxim announced the

invention, early in 1908, it created a sensation. Apparently to throw patent snatchers off the trail, he informed reporters that the silencing was done by a little valve within the tube, which swung shut across the rifle barrel immediately after the bullet had passed, and before the gasses arrived. Quite a device! He stuck to the story and laymen took it seriously for a whole year before Mr. Maxim decided there was no longer any need for keeping the mechanism a secret.

"Terrifying Possibilities" Seen

New York papers, on the whole, hailed the Maxim silencer as "terrifying in its vast possibilities." Honest citizens had nightmares in which footpads crept about exterminating whole neighborhoods, while policemen stood at ease on the sidewalk and heard no warning sounds. The silencer provided material for a generation of mystery novelists and playwrights. To this day, the average man thinks the silencer works perfectly on a pistol, although Mr. Maxim was giving demonstrations a quarter of a century ago to prove that such was not the case.

The Maxim silencer really works well only on a breech-sealed rifle, where all gas and noise must pass through the barrel. The revolver and the automatic pistol do not possess this feature — the revolver having a necessarily loose joint between the cartridge cylinder and the breech end of the barrel, while the automatic has its recoil and reload mechanism as a source of noise distribution.

A properly constructed rifle, fired with the silencer, gave only a quiet metallic "plop," Mr. Maxim showed. A pistol fired with the device gave a report only slightly softer than the sound produced unsilenced.

Mr. Maxim later devised airplane parts and perfected relay systems for transmitting telegraph and telephone messages without loss of power. In 1832 he caught the public eye in no uncertain fashion by announcing that the Maxim Silencer Company was going to start a consulting and designing service in the elimination of the noises of large cities. He even published a fantastic trademark, "Dr. Shush," a little, bewhiskered chap with monstrous spectacles and a hushing finger held to his lips.

Designed "Sound Conditioned" Rooms

He designed his "sound conditioned" rooms by closing direct ventilation openings and inserting in them modified Maxim silencers equipped with electric exhaust and intake fans. He was aided in this work by his son, Hiram. Mr. Maxim had married Miss Josephine Hamilton, daughter of a former Governor of Maryland, in 1898.

It was the Maxim company's motto that any noise emerging from a pipe could be silenced effectively, economically and with no significant loss of mechanical efficiency.

Seven years ago Mr. Maxim took to writing on scientific subjects with a sprightly, amusing style. In 1933 his book, "Life's Place in the Cosmos" was published, in which he said that on the outermost planets life once existed but had since been frozen into extinction.

The innermost planets, in his opinion, were too hot to support life at the present time, though life might be found on them in the future. The existence of life on those planets, when the time came, he wrote, might be verified through radio communication.

Last fall he began publication in "Harper's Magazine" of a series of articles about his father. Written somewhat in the manner of the stories of the late Clarence Day, Mr. Maxim's stories nevertheless retained a full flavor of the nineteenth century. He also was the author of a syndicated column in the Hearst newspapers.

Experimented With Radio

The last ten years of his life brought him a great deal of pleasure and amusement in his work with amateur radio transmission and motion picture photography. He was president of the International Amateur Radio Union and of the Amateur Cinema League. He was a member of the Republican Party and of the Unitarian Church, and held the rank of Lieutenant Commander in the United States Naval Reserve. He was a member of the American Society of Mechanical Engineers, the Society of Automotive Engineers, and of so many technical and scientific clubs that he never could or would list them all for publication in works of biographical reference. Colgate University conferred the degree Doctor of Science upon him. He held membership in the Engineers' Club of New York and in the Engineers, Automobile, Aero and Golf Clubs of Hartford.

In his lifetime, perhaps, Hiram Percy Maxim did not surpass the world fame of his father and uncle. But if batteries of Maxim silencers eventually achieve a diminution of the nerve-torturing noises of mankind's cacophonous cities, then Hiram Percy's name may be blessed by generations of the future.

Reprinted from *The New York Herald Tribune,* February 18, 1936.

The following press release was written by Byron H. Goodman, W1DX, who retired from HQ in 1965 after 32 years service. Release date February 16, 1936.

Amateur Radio News from the
AMERICAN RADIO RELAY LEAGUE
West Hartford, Connecticut.

For Release Friday

West Hartford, Conn., Feb. 21—Amateur radio stations will today observe a nation-wide silent period as a mark of respect for their late beloved leader, Hiram Percy Maxim.

As the famous scientist and inventor is being interred at Hagerstown, Md., thousands of radio amateurs throughout the country will silence their transmitters in tribute to the man who has been their friend and champion during the past two decades.

Interested in amateur radio since 1907, Maxim was famous in amateur circles as the founder and president of the American Radio Relay League, national amateur body organized in 1914, and of the International Amateur Radio Union, federation of international amateur radio societies organized in 1925. A strong believer in the rights of radio amateurs, he has appeared several times before Congressional committees and at international radio conferences, championing the cause of the amateur and protesting against possible unfair legislation.

When in 1914 Maxim realized the need for organization of the hundreds of radio amateurs scattered throughout the country, he hit upon the idea of the relaying of radio messages as the common bond between amateurs, hence the name "American Radio Relay League". He has been the only president of the the League since its beginning, and although he several times voiced the opinion that another man should occupy the position, the amateurs were insistent on his remaining in the chair. The same traffic networks that Maxim played such a large part in organizing kept the newspapers of his home town informed of his condition during the siege of illness that overtook him at La Junta, Colorado, while en route to Lowell Observatory in Arizona, and ultimately flashed the sad news of his death to amateurs throughout the world.

Today these networks are stilled, in silent tribute to his memory.

BHG: 2-19-36-D14

WESTERN UNION

THE COMPANY WILL APPRECIATE SUGGESTIONS FROM ITS PATRONS CONCERNING ITS SERVICE

1201-S

CLASS OF SERVICE

This is a full-rate Telegram or Cablegram unless its deferred character is indicated by a suitable symbol above or preceding the address.

R. B. WHITE, PRESIDENT
NEWCOMB CARLTON, CHAIRMAN OF THE BOARD
J. C. WILLEVER, FIRST VICE-PRESIDENT

SYMBOLS

DL = Day Letter
NM = Night Message
NL = Night Letter
LC = Deferred Cable
NLT = Cable Night Letter
Ship Radiogram

The filing time shown in the date line on telegrams and day letters is STANDARD TIME at point of origin. Time of receipt is STANDARD TIME at point of destination.

Received at 708 14th St., N. W. Washington, D. C.

```
WAA46 11 XU=WUX HARTFORD CONN FEB 17 1936 1127A

K B WARNER=
```

```
      HOTEL SHOREHAM=

UNITED PRESS REPORTS MR MAXIM DIED
THIS MORNING STOP ANY INSTRUCTIONS
QUERY PLEASE ACKNOWLEDGE=

      BUDLONG.

         1212P.
```

NO ADDITIONAL CHARGE IS MADE FOR REQUESTING A R

The Reason Why

By Hiram Percy Maxim, President A.R.R.L.

(Reprinted from *QST* for September, 1927)

SITTING back in the old armchair, with the last issue of *QST* read from cover to cover and with everybody else in the house asleep hours ago, I fell to thinking of amateur radio to-day and amateur radio of other days. As the blue smoke curls slowly upward from the old pipe, visions of early A.R.R.L. Directors' Meetings float before me. I see those old-timers grappling with problems of organization, with QRM, with trunk-line traffic and rival amateur leagues. I see sinister commercial and government interests at work seeking to exterminate amateur radio. They were dark days, those early ones.

To-day I see Amateur Radio an institution, recognized by our American government and on the road to recognition by the other governments of the world. I see a fine, loyal A.R.R.L. membership of 20,000 standing shoulder to shoulder and believing in each other and still blazing the way in radio communication. I see a rapidly developing world-wide amateur radio brotherhood taking shape, in the form of our I.A.R.U.

And as the last embers of the old pipe turn to grey ash, I ask how it all came about: that the A.R.R.L. should have succeeded and all its opponents failed. The answer is clear. It is because with our opponents there was always some kind of a selfish motive to be served for someone, whereas in our A.R.R.L. we insisted from the beginning that no selfish motive for anybody or anything should ever prevail. Everything that A.R.R.L. undertakes must be 100% for the general good. That policy bred loyalty and confidence. With those two things an organization can prosper forever.

W1AW—Amateur Radio's "Dream Station"

1936 began as a year of tragedies for the American Radio Relay League. In February, Hiram Percy Maxim died unexpectedly. A month later, a devastating flood hit the northeast: Hartford was one of its principal victims. The ARRL HQ station at the time was W1MK, located at Brainard Field in Hartford. W1MK was completely inundated by the flood waters. As though a thick coating of silt over the station and everything in it were not enough, several tanks of fuel oil burst and seeped into the building. The only items saved were those that could be carried away by the operator, such as the log books and other records.

In May, the ARRL Board of Directors decided to erect a new Headquarters station. W1MK schedules were being kept by station W1INF at the ARRL Headquarters building in West Hartford. ARRL applied to the FCC for, and received, Mr. Maxim's personal call, W1AW, to be used at the Headquarters station as a permanent memorial to HPM. On February 17, 1937, the first anniversary of HPM's death, W1AW made its inaugural appearance on the air as part of an operating activity called the Maxim Memorial Relay.

Plans began in earnest for the construction of the "dream" station. In late summer, 1937, a seven-acre site in then relatively open country in Newington (now a densely populated Hartford suburb) was purchased as the home for the new installation. This location was just four miles south of the ARRL Headquarters offices in West Hartford. "When completed, the station will have four completely separate transmitters, with full amateur power capability for each amateur band, and provision for radiotelephone and radiotelegraph work included for each. It will be a station of which any member may well be proud," announced March 1938 *QST*.

Finally, on the afternoon of July 9, 1938, the big moment arrived. Chief Operator Hal Bubb, W1JTD, observed by ARRL officials and leading local amateurs, called CQ on 3520 kHz and was promptly answered by W2LC on Long Island. In all, 19 contacts were made during the inaugural operation.

According to a description that appeared in *QST*, the new station was "sort of an amateur's dream." There was a one-kilowatt transmitter for each amateur band from 160 to 10 meters. Each was designed for break-in operation on CW. For phone operation, there was a common modulator which could be switched manually to any of the transmitters. The receiver was an HRO-5, a real classic of its day. The HRO and a Boehme tape keying head were probably the only two pieces of commercially built equipment in the station.

If all this seems Spartan for a "dream" station, remember that Amateur Radio in 1938 was vastly different than it is at present. Almost all the equipment was home built. Low power was the order of the day, with most amateur transmitters in the 50-watt or less range, and CW only. Kilowatt transmitters, especially on phone, were extremely scarce and worth traveling miles to observe and operate.

The antenna farm was equally impressive. The *piece de resistance* was the rhombic antenna, 350 feet on each side of the diamond, 55 feet up, and aimed due west. It was unterminated, so it gave excellent bidirectional coverage. A two-wavelength Hertz was used on 40 meters, a half-wave Hertz on 80, and another half-wave Hertz on 160. The rhombic, while designed for 20 meters, was equally at home on 10, 40, or 80 meters.

Two regular operators were assigned the job of maintaining W1AW and keeping all the

(W2ABE photo)

The first ARRL Headquarters station! The transmitter was four UV202s in parallel in a self-excited Hartley circuit. The so-called "five watters" mustered an input power of 20 watts total. The receiver, on the left, was the familiar UV200 detector and one step of audio amplification. Note the battery power supply. The receiver was surprisingly sensitive, but not very selective!

W1MK's transmitter before the flood of 1936. The flood damage spelled the end of an era for the ARRL Headquarters station at its Brainard Field, Hartford, location. Activity was transferred to 38 LaSalle Road, West Hartford, first as W1INF, then W1AW until the new Maxim Memorial Station was erected.

regular schedules. In addition to overseeing the CW and voice bulletin transmissions, the operators kept many traffic schedules, both in nets and with individual stations, and contacted as many other amateurs as possible as time permitted.

On September 2, 1938, which would have been HPM's 69th birthday, the formal dedication was held. ARRL President Eugene C. Woodruff, W8CMP, unveiled the Maxim Memorial Tablet in the presence of a large number of League members, all ARRL officers and Headquarters staff, and dignitaries from the Town of Newington and State of Connecticut. The ceremonies were carried live on the CBS Radio Network.

W1AW was quite active in the events surrounding the entry of the US into World War II. Although regular amateur operation was shut down by the FCC, the message first went out via W1AW at 9:59 PM EST on December 7, 1941. Many amateurs did not get the word, so W1AW was authorized to contact these stations and explain the situation. W1AW then continued to transmit bulletins, many of them recruiting amateurs—and their equipment—for the war effort. Finally, on January 10, 1942, the station made its final transmission and left the air for the duration, with this final log entry: "73 until we meet again!!"

W1INF. This station started signing W1AW in February 1937, when the FCC took special action to assign Hiram Percy Maxim's old call to ARRL.

The W1AW property immediately after its purchase in 1937. The pavement in the foreground is Main Steet in Newington. The automobile is parked on what is now Starr Avenue, marking the southern boundary of the property.

Work on the new station progresses quickly.

And we did meet again! The station was cleaned up, overhauled and rarin' to go. On October 31, 1945, the first postwar bulletin was sent. Although probably no one realized it at the time, the halcyon days of Amateur Radio were about to end, and a period of change in regulations, equipment and operating practices was to begin.

By May 1946, enough of the bands had been returned to amateur use to allow the postwar resumption of what has probably been W1AW's greatest service to the amateur fraternity, the ARRL Code Proficiency Program. The code-practice transmissions resumed with text from 15 to 35 WPM. In 1951, the FCC created the Novice license with a code speed requirement of 5 WPM. W1AW changed its program format accordingly. Eventually, the present schedule of four hours each weekday and three hours a day on weekends evolved.

W1AW transmitters in the making. This photo shows the 3.5 MHz rack with power supply, 89-6L6-RK48 driver, and modulator (T-822s) at right. Switches connect this 500-watt modulator to "any" transmitter. Push-pull KH354s (above) are used in the final amplifier.

Spring 1938. The station is nearly completed. At this point in history there were 48,000 licensed US amateurs.

The dedication of W1AW, September 2, 1938. The ceremony was carried by WTIC and WDRC locally and by CBS nationally. George Bailey, W1KH, ARRL's vice president, was master of ceremonies.

In the early 1950s, what was probably the most important technological breakthrough ever, or at least since CW supplanted spark, began to revolutionize Amateur Radio. Single sideband (SSB) was on its way to becoming the standard mode for voice transmission. In addition to providing a significant improvement in equipment efficiency, the popularity of SSB had much to do with the development of the transceiver and the linear amplifier in amateur operations. W1AW began SSB operation by making contacts with others using the mode. As its popularity increased, SSB bulletin transmissions began, first on one or two bands and finally on all bands.

In August 1955, the emergency preparedness of W1AW was tested when severe flash flooding hit the northeast. With the entire HQ staff pitching in to help, W1AW was on the air more than 20 hours a day for ten days, first handling emergency traffic and later health and welfare messages.

Another new mode was radioteletype. RTTY was initially permitted on the 11-meter band only (then an amateur allocation). But when the FCC authorized RTTY operation on all HF bands, W1AW incorporated an RTTY bulletin program. Much later, when enhanced digital modes ASCII and AMTOR were authorized by the FCC in 1980 and 1983 respectively, they were immediately included in the W1AW teleprinter program. Incidentally, W1AW was the eastern end of the first amateur coast-to-coast RTTY contact.

An inspiring incident of this era was the opening of the 7200-7300 kHz portion of the 40-meter band to phone operation, to take effect at 3 AM EST on February 20, 1953. At 3:00 on the dot, there must have been 5,000 stations, W1AW included, from all over the country all calling CQ at the same time! Later, W1AW also participated in the "bandwarming" for the new 10.1, 24 and 18-MHz bands in 1982, 1985 and 1989, respectively.

Another interesting event was the initial pass of the first OSCAR (Orbital Satellite Carrying Amateur Radio) over the Connecticut area on December 12, 1961. Several members of the HQ staff joined the W1AW crew to listen for the satellite. As the appointed moment approached, you could have heard the proverbial pin drop.

W1AW operating room just after World War II.

The waiting room at W1AW, where visitors could wait until the operator had a chance to show them around. That's "Old Betsy," Hiram Percy Maxim's spark transmitter, in the corner. This waiting room went the way of the workshop and the front lobby when the building was renovated in the 60s.

Then OSCAR came into range and its cheery "HI" was heard for the first time.

During the early 1960s, the new ARRL Headquarters building was erected in Newington, right in the middle of the W1AW antenna farm. After the new HQ building was completed, it was decided that W1AW should be completely remodeled. A new approach to simultaneous transmissions was installed—a common exciter, operating at 3 MHz, driving a string of transverters that converted the 3-MHz signal to each of the various operating frequencies. A driver and kilowatt amplifier for each band completed the transmitter chain. The driver/amplifiers were designed and built by Bill Orr, W6SAI, and the ARRL HQ lab built the rest of the equipment. Additionally, a visitors' operating position was set up, and the W1AW tape-generated CW system was updated. Instead of Wheatstone tape, regular five-hole typewriter tape and a commercial teletypewriter-to-Morse converter was used.

As newer OSCAR satellites carried transponders for amateur use, W1AW installed equipment to utilize this new mode. Hundreds of contacts were made, and are still being made, through the OSCAR series of satellites and, later, the Soviet Union's RS satellites.

W1AW entered the computer era with a Heath H89 system. The code practice transmissions were counter generated, followed by the Baudot, ASCII and AMTOR transmissions. For CW, speeds of 5 to 70 WPM were available.

W1AW has had several high points in recent years. W1AW contacted Astronaut Owen Garriott, W5LFL, in the space shuttle *Columbia*. The station played an important public service role during the US invasion of Grenada, when the only information source was the ham radio transmissions of one of the medical students. W1AW also participated in

W1AW operating room just prior to renovation in the early 60s.

W1AW as it appeared after renovation in the late 60s.

the communications and relief efforts following the devastating Mexican earthquake. In 1989, W1AW played a key emergency communications role when Hurricane Hugo devastated the Caribbean islands including Puerto Rico and the US Virgin Islands. Closer to home, W1AW was staffed and ready for action during Hurricane Gloria in 1985, and the station operated on emergency generator power for more than 48 hours following the storm.

The electronics field changes rapidly, and W1AW underwent renovation to meet challenges of the '90s and the 21st century. On July 20, 1989, the most recent rededication ceremony took place for a refurbished, modernized W1AW. ARRL President Larry E. Price, W4RA, presided over the event and cut the ribbon with ARRL Executive Vice President David Sumner, K1ZZ. Many ARRL officers, directors, League staff and members attended the event.

Today, W1AW uses seven identical commercial-grade transmitters for code practice and bulletins. These all-solid-state transmitters are capable of running one kilowatt output continuously. In addition to the code practice/bulletin transmitters, W1AW is a showcase for the latest amateur gear. Three visitor operating positions featuring equipment loaned by many amateur manufacturers are available for general operating on 160 through 10 meters, as well as VHF/UHF and satellites. The four towers are loaded with antennas for all bands from 1.8 through 1296 MHz. Beams for code practice/bulletin transmissions are fixed west, and separate rotatable antennas are available for general two-way operation.

As technology progresses, W1AW will keep in step. Every generation of amateurs will be able to picture W1AW as their "dream station," and take great pride in its accomplishments. (By Charles R. Bender, W1WPR, Chief Operator, W1AW. Excerpts reprinted from *Popular Communications* with permission.)

"Help! I'm being held prisoner in W1AW's transmitter!" So says famous world-traveler DXer extraordinaire, Gus, W4BPD, during a spring 1966 visit to W1AW and ARRL Headquarters. Guarding the rack are W7PHO (left) and W2GHK (right). That's W1BDI looking over Gus's shoulder.

"Carrying on the tradition of several decades, the Hiram Percy Maxim Memorial Station W1AW continues to provide an important service to radio amateurs everywhere with code practice, bulletins, oscar orbital information and other valuable transmissions. Today's increased interest in amateur radio requires that this dedicated service be continued and plans for future expanded operations be made. ARRL pledges its active support of this operation. Further, as the date of the World Administrative Radio Conference approaches, the ARRL speaks with its radio voice asking that radio amateurs worldwide continue to hold the extension of international goodwill via amateur radio as one of our most cherished goals. Please join me in wishing a hearty happy birthday to W1AW. 73."

**February 17
1937-1977**

Harry J. Dannals, W2HD
President, ARRL

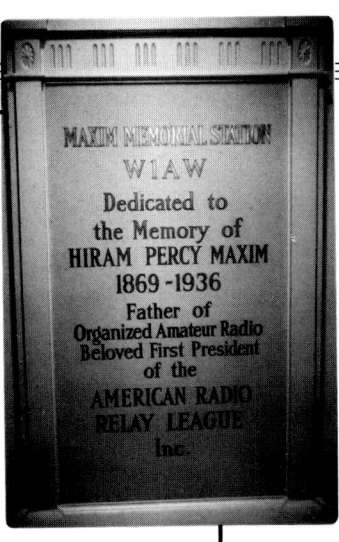

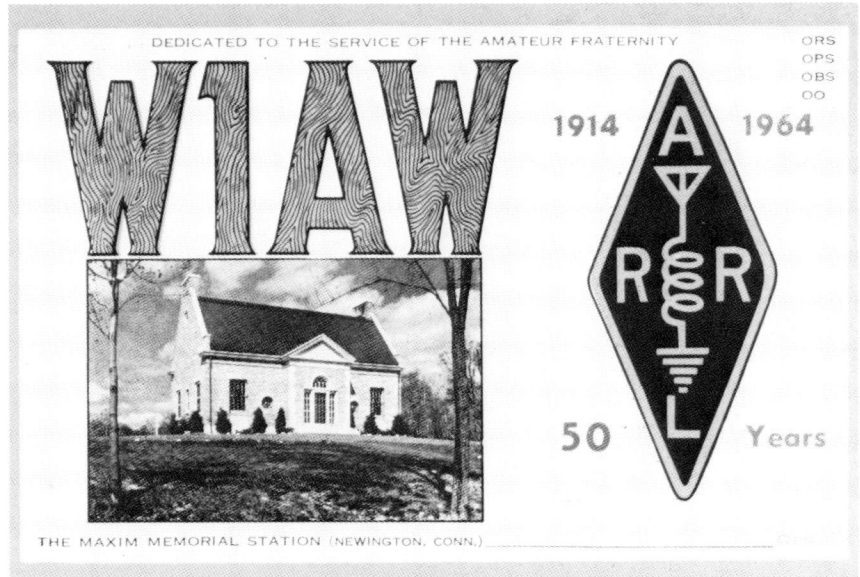

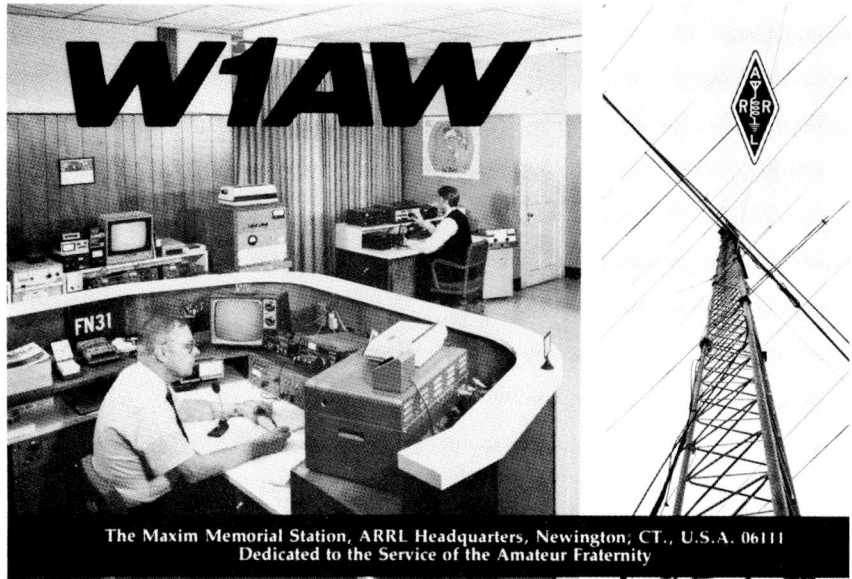

W1AW—ANOTHER NEW FACE

Tod Olson, KØTO, ARRL Vice President of International Affairs, used one of W1AW's guest operating positions to make some of the first contacts from the renovated station.

More than 200 people gathered at the corner of Main Street and Starr Avenue in Newington to witness the unveiling of the renovated W1AW on July 20, 1989. Under a huge canopy, the crowd listened to Amateur Radio and local dignitaries extol the virtues of the famous little brick building and the services it provides. American Red Cross External Relations Manager Bobby Baines commended the League's years of service with the Red Cross. Seated are (right) Larry E. Price, W4RA, ARRL President and David Sumner, K1ZZ, Executive Vice President.

W1AW Chief Operator Charles Bender, W1WPR, surveys his new surroundings from the main operating console.

A big part of the 1989 W1AW renovation is a new antenna system featuring multiple beams for 40 through 10 meters, wire antennas for the low bands, and arrays for VHF, UHF and satellite work. Mark Wilson, AA2Z, (left) assembles one of the 20-meter bulletin Yagis designed by Bill Myers, K1GQ, and manufactured by Cushcraft. Jeff Bauer, WA1MBK, installs the first of many antennas on the 120-foot tower.

Fred Hammond, VE3HC, donated the equipment racks that house the 1-kW solid-state Harris bulletin transmitters. W1AW can transmit bulletins and code practice on up to seven HF/MF bands simultaneously.

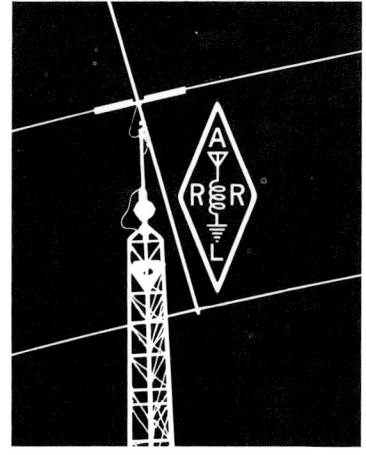

CARTOONS THROUGH THE YEARS

Vacationer's First Aid

Better yet... Switch to Safety

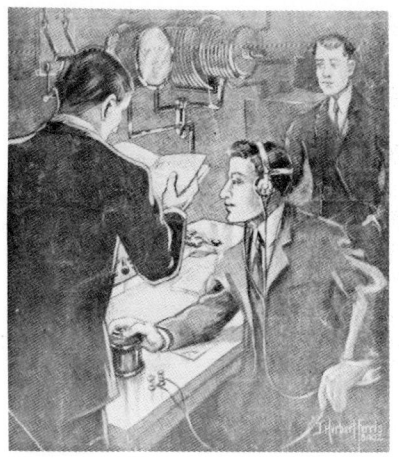

Strays

How's DX?

ARRL...The Dream—The Reality

Hiram Percy Maxim's dream was the creation of a national association of amateur operators, joined together to relay messages beyond the range of an individual station.

In 1914, that dream began to take form in Hartford, Connecticut. On January 14th, Hiram Percy Maxim chaired the first meeting of the Radio Club of Hartford. He had become interested in Amateur Radio through the activities of his son, Hiram Hamilton Maxim, and soon developed one of the dominant stations of all New England.

The relay idea represented an ideal basis for needed national organization. Some basic principle, some prime moving force, was essential for the success of such an organization. Americans had always been great joiners, but if an amateur organization were ever to progress beyond the paper stage, or expand into more than a local club, it must offer more than a gaudy membership certificate and one's name on the rolls.

At best, radio ranges in those days were limited. With the power, equipment, and wavelengths available, there was little hope for greater distances. After all, the only way amateurs of those days knew how to reach greater distances was to increase power. Unfortunately, they were limited to one kilowatt. Even if this were stretched to two or three, as was occasionally done, the improvement was not appreciable. But an intermediate amateur could relay messages over greater distances with ease and expedition. The only requirement was to achieve some sort of mutual understanding so that each amateur would aid his fellows. Organization was needed —organization that would accomplish the dual purposes of opening relay facilities to all, and of bonding together the amateurs of the country into one strong, cohesive, self-reliant body.

Mr. Maxim discussed his idea with the Hartford Club president, David Moore, and then wrote him the historic letter reproduced in these pages. At its meeting on April 6, 1914, the Radio Club of Hartford voted to take charge of the development of a relay organization, and a committee appointed to handle the details.

The League Grows

Maxim and Clarence Tuska set up shop in the Tuska family home attic. By mid-May, President Maxim and Secretary Tuska developed

W1AW stood alone on the Main Street, Newington, property until 1962, when ground was broken for a new Headquarters building.

ANNOUNCEMENT

¶ Q S T is published by and at the expense of Hiram Percy Maxim and Clarence D. Tuska.

¶ Its object is to help maintain the organization of the American Radio Relay League and to keep the Amateur Wireless Operators of the country in constant touch with each other.

¶ Every Amateur will help himself and help his fellows by sending in 25 cents for a three months' trial subscription.

THE PUBLISHERS OF Q S T

Clarence Tuska, cofounder of the American Radio Relay League.

The League's home from 1919 to 1922 at 721 Main Street, Hartford. *(R. A. Stevens photo)*

Twelve staffers manned our new quarters at 1045 Main Street, Hartford. HQ occupied three rooms on the third floor.

Staffers at ARRL Headquarters at 1045 Main Street, Hartford, in 1923.

application blanks announcing the formation of the American Radio Relay League and sent them to every amateur station they could think of. There were no dues; membership was free upon application. At the same time, the requirements were set at a high standard and rigidly maintained. Only qualified amateurs were accepted as relay stations. The response was tremendous. The influence of the League was mounting rapidly. It had members in every section of the country...

By August 1914, more than two hundred relay stations had been appointed, from Maine to Minneapolis and from Seattle to Idaho.

In late 1914, Maxim went to Washington and conferred with the Commissioner of Navigation of the Department of Commerce. The object of the conference was to establish the League in official circles, and to secure the important concession of permission to operate stations at strategic points along the relay routes of the country under restricted special licenses, enabling them to use the wavelength of 425 meters. These licenses were issued wherever necessary to enable relaying to the next point on the chain, and they were granted only to stations sufficiently remote from the seacoast to avoid interference. The sole restriction was that the 425-meter wavelength was to be used exclusively for the relaying of bona fide messages, and not for idle conversation.

The League was expanding rapidly and was forced to give careful consideration to the question of finances. Members were asked for a nominal contribution to help defray expenses. It was strictly voluntary. There was, however, a gentle hint that non-paid-up members would be so listed in succeeding issues of the call-book.

The membership grew steadily. Amateur interest and proficiency did also. In early 1915, the Hartford Radio Club and the League came to a parting of the ways. By the end of 1915, amateur stations were accomplishing what were in those days unbelievable feats in transmission and reception. With homemade equipment, often not exceeding a hundred dollars in total cost, and in the despised 200-meter region, they were frequently outperforming government and commercial plants representing investments of thousands of dollars.

It was becoming an insurmountable task for the League to acquaint the growing membership with new plans and schedules by means of correspondence alone. It became increasingly apparent that some kind of general circular or bulletin was necessary. The League, however, had no funds.

The answer, seemingly obvious but surveyed with some reluctance by Maxim and Tuska, was a self-supporting magazine. In December 1915, each member of the League received in his mail a sixteen-page magazine called *QST*—the "December Radio Relay Bulletin." This, it was announced, was being published privately at the expense of Maxim and Tuska. It was to be sold independently of the League, on a subscription basis. The subscription fee was to be $1.00 per year. The stated object of the magazine was "to maintain the organization of the American Radio Relay League and to keep the amateur wireless operators of the country in constant touch with each other."

QST first appeared in December 1915, financed by Maxim and Tuska. It had an immediate beneficial effect: Membership jumped sharply, from 635 on December 1, 1915 to 961 on January 10, 1916. It also served, more effectively than before, to inform members— and posterity—of what the League was doing.

The month of February 1917 is of historic importance in Amateur Radio. That month marked the beginning of a major change in the governing structure of the ARRL. For nearly three years, Maxim and Tuska, serving as

ARRL Headquarters at 1711 Park Street. There wasn't much room, so the transmitters were allotted space in the Circulation Department. The location was such that visitors often yielded to the temptation of twisting dials, throwing the equipment out of adjustment. The transmitter received unintentional jolts from the nearby addressing machinery, and the operating desk was in a conference room somewhat removed from the transmitter. When the keying relay stuck, the operator would frequently be seen dashing wildly down the corridor to Circulation to correct the problem.

38 LaSalle Road, West Hartford.

president and secretary respectively, had been the sole officers of the League. By 1917, it had reached such size and importance that a more suitable organization was deemed advisable. Consequently, on February 28, 1917, a group of leading amateurs met at the Engineers' club in New York City to consider the problem. After a succession of meetings, they had written and adopted a constitution that outlined the policies of the League, specified the machinery for the election of officers, divided the country into six divisions, elected by vote twelve ARRL directors and four officers, and declared membership open to anyone interested in radiotelegraphy or radiotelephony...

War

In April 1917, all licensed amateurs received the following letter from the office of the Chief Radio Inspector of the Department of Commerce:

"To all Radio Experimenters,

"Sirs:

"By virtue of the authority given the President of the United States by an Act of Congress, approved August 13, 1912, entitled, 'An Act to Regulate Radio Communication,' and of all other authority vested in him, and in pursuance of an order issued by the President of the United States, I hereby direct the immediate closing of all stations for radio communications, both transmitting and receiving, owned or operated by you. In order fully to carry this order into effect, I direct that the antennae and all aerial wires be immediately lowered to the ground, and that all radio apparatus both for transmitting and receiving be disconnected from both the antennae and ground circuits and that it otherwise be rendered inoperative both for transmitting and receiving any radio messages or signals, and that it so remain until this order is revoked.

Breaking ground—1962

Aerial View—1963

Immediate compliance with this order is insisted upon and will be strictly enforced. Please report on the enclosed blank your compliance with this order; a failure to return such blanks promptly will lead to a rigid investigation.

Lieutenant, U.S. Navy,

District Communication Superintendent."

Immediately following this crushing blow, Amateur Radio was called upon to defend itself from a legislative menace. The Padgett Bill, H.R. 2753, introduced in the House on April 9, 1917, proposed that all radio communications in the United States, including amateur, commercial, and extra-Naval governmental stations, be turned over to the Navy.

Naturally, all the radio world rose in protest. Individual amateurs generally disapproved the bill in principle, even though none of them dared say when they would actually be allowed to operate stations again. Charles H. Stewart, representing the Wireless Association of Pennsylvania and a number of other clubs, was heard in protest during the House Committee hearings. The National Amateur Wireless Association, through *The Wireless Age*, fought the measure bitterly. Hiram Percy Maxim, representing the ARRL, went to Washington to confer with the sponsors of the bill, and secured an exception from its provisions for amateur stations, if and when they should be permitted to reopen. The bill was eventually killed in committee, but the incident is of historical significance in that it showed that even at this early date the ARRL was accepted as the organization which represented Amateur Radio. Its membership total of about 4,000 was not as high as that claimed by competitive organizations, but by far the greatest percentage of licensed amateurs was enrolled among its numbers.

That threat over, Amateur Radio settled down to its next job, helping Uncle Sam to win the war.

There can be no question of the importance of the part that radio amateurs played in the winning of the war. The superiority of Allied, and particularly American, communications was the deciding factor in many moments of close struggle during the fighting on all fronts.

The early post-war years brought several changes. In 1917, a formal constitution was adopted by the Board of Directors. The Board members were: HPM, Clarence Tuska, and Attorney Lawrence A. Howard. In 1919, Kenneth B. Warner was hired as Secretary. The

Early view of the finished headquarters building.

Headquarters of the American Radio Relay League, Inc, as it appears today.

League purchased *QST* rights from Clarence Tuska and became the official publisher of *QST*. More working space was needed, and office space was rented at 721 Main Street in Hartford.

In 1922, when the League staff had grown to 12 and the League had 7,400 members, Headquarters moved to 1045 Main Street in Hartford.

Three years later, with a staff of 25 and 19,000 members, the office was moved to an entire floor of 1711 Park Street, Hartford.

In 1931, the League occupied the second floor of 38 LaSalle Road, West Hartford. The building was brand new, and the office space had been laid out by the League staff. The new office was referred to in the April, 1931 issue of *QST* as "...a quieter location..."—but there was a bowling alley in the basement and plans for an indoor miniature golf course on the first floor!

In 1937, the staff had grown to 36, the membership to 23,000, so Headquarters took over the ground floor for circulation and shipping operations.

The bowling alley moved out in the late thirties, and in 1945 the lab, mailroom and storage rooms were moved to the basement, completing ARRL occupancy of the building. At that time, there were 47 employees and 42,000 members.

From 1945 to 1963, membership grew to over 105,000. Thanks to donations from many members, a new League Headquarters was built at 225 Main Street, Newington, right behind W1AW. This was the first time in many years that Headquarters had adequate space to administer the League's many programs.

By 1977, membership grew to over 143,000 and Headquarters expanded to handle the increase in membership.

Today, we serve over 160,000 members worldwide from our Newington Headquarters. Over the years, technology has changed, and our locations have been many. But the following text, written by Hiram Percy Maxim in 1919, timelessly explains our mission.

The Importance of our ARRL

...There is no such thing as organization if each one of us starts out to be thoroughly selfish. If all our efforts are to be directed solely for the benefit of self, we are purely individual and able to take about as much form as the individual sands of the sea. We are an incoherent, uncontrolled crowd. On the other hand, if just a little of our efforts is devoted to the common cause, we automatically establish organization and efficiency and protection, and everything else that is elevating, improving and worth the having...when an amateur asks that old-time question, "what do I get out of joining the ARRL?" the answer should be, "Protection." He cannot have it unless somebody joins an organization and does the work. Unless he joins and does his bit, he must not complain when his fellows place him in the list of unenviables who are not willing to do any work themselves...
—Hiram Percy Maxim

David L. Moore, President,
Radio Club of Hartford,
18 Asylum Street,
Hartford, Conn.

My dear Mr. Moore:—

I am enclosing herewith a copy of letter which I have sent to *Modern Electrics* and also to *The Electrical World*. As you will see it "opens the ball" on the subject of our Relay Scheme.

Now what I want to do is to get you and Tuska together some time, within the next day or two, and organize the AMERICAN AMATEUR RADIO LEAGUE. We three can draw up in a few minutes a very simple straight forward statement of the objects of this League. We can then decide who the officers should be and elect them. Then, at the next meeting of the Radio Club of Hartford, we can let the Club decide if it is to become a member of the League. We will then be regularly started and can probably get the Connecticut Valley Radio Club in Springfield to join and it would not be long before we could get others also.

The object of securing the membership of the various Clubs, would be to have those Clubs advise us as to what stations in their locality are the best ones for us to appoint as OFFICIAL RELAY STATIONS. We probably would get wise advice in this manner, because it would be quite a distinction for a station to be appointed to a long distance relay point. It is the only way we will have of getting at the proper stations who could be counted upon to always be in working order and able to read and transmit at decent speeds.

My letter describes the whole matter. I am sending a copy of this letter to Tuska. I wish both of you would give this subject careful thought and be prepared to bring up all possible objections so that we will make no mistakes in the beginning.

Very truly,

Hiram Percy Maxim, W1AW

Hiram Percy Maxim

Your League: Serving Amateur Radio for 75 years

Upper left:

This 1916 cartoon still applies!

Upper right:

(Left to right) Lt. Follett Bradley, Observer and Radio Operator and Lt. Henry H. Arnold, pilot, in the world's first successful radio air-to-ground test at Ft. Riley, Kansas, on November 2, 1912. *(Official photograph, US Army Corps)*

Lower left:

(Left to right) Lt. Joseph O. Bauborgne and Lt. Follett Bradley who installed and operated the first military air radio set at Ft. Riley, Kansas, on November 2, 1912. *(Official photograph, US Army Air Corps)*

Lower right:

The goal of every amateur was a bigger and better antenna system. This 80-foot-high antenna, located at Prospect Avenue in Hartford, belonged to Hiram Percy Maxim.

December 1922—C. W. has resoundingly passed spark as the mainstay of message traffic nets, by nearly a 4 to 1 ratio; earlier in the year, it was exactly the reverse. Totals of both modes reach a peak figure of 30,000 messages a month in winter.

36

UNCLE SAM: "I RECKON MY AMATEURS LEAD THE WORLD."

June 1920—There are now twelve instead of six divisions in the League, resulting in a less burdensome workload for the various Directors.

Stations of this early era were a blend of home-brew and purchased equipment. The emphasis, however, was very heavy on home-brew.

June 1924—A Senate bill to levy a 10% excise tax on all radio gear was publicized through League channels and the resulting flood of protests to Congress laid the matter to rest.

December 1922—Northwestern Division Manager Howard F. Mason, 7BK describes "The Radio Lizz," a flivver equipped for reception (detector and two-step) with 78 feet of wire in a three-turn loop completely around the car—enough to frighten the most docile passing horse! He says ignition interference is nigh insurmountable.

December 1924—An evaluation of commercially built air-core transformers will whet the interest of many hams seeking the better performance a superhet will provide.

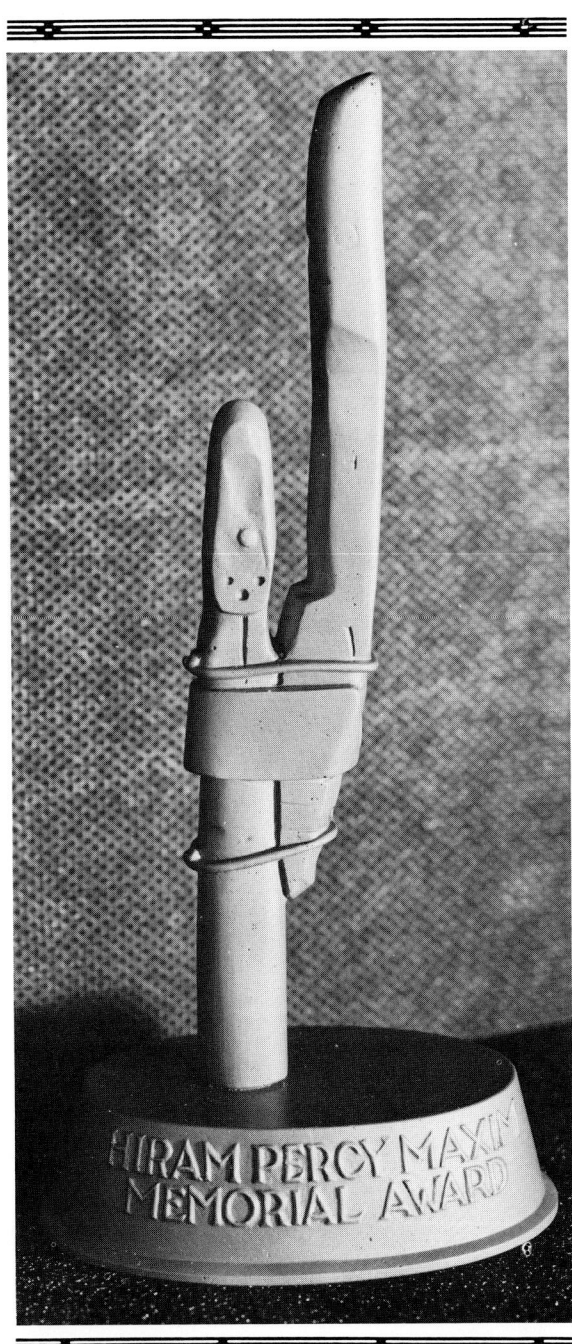

The Wouff Hong

In an institution as old as Amateur Radio, traditions and symbols of the art appear and become part of it. Our traditions are many, but aside from the ARRL diamond, only one has become a part of the framework of Amateur Radio, the symbol of its finest traditions, its long and glorious history.

That symbol is the Wouff Hong. Every ham should know its origin.

It started back in 1917, in the very earliest days of ARRL and *QST*, when an anonymous amateur, writing under the title "The Old Man," created a wonderful series of humorous stories in the magazine. In a pithy, irascible style he assailed all that struck him as criticizable about ham radio operation of the period in his famous "Rotten Radio" series. He pitilessly exposed the poor operating practices of the day, yet did it in a way which drew chuckles even from those recognizing themselves as the special targets of his ire.

In one of those stories, "Rotten QRM," he launched forth with examples of some of the poor sending cluttering up the band in a particular QSO to which he was listening. The gibberish included the words "wouff hong" which, apparently, was being used by someone on somebody else.

Although T.O.M. himself admitted at the time he didn't know exactly what a wouff hong was, it quickly became something with which both to attack bad operating practices and to discipline their perpetrators.

The tradition was established, and the Wouff Hong created in the minds of thousands of amateurs as some mythical instrument of torture to be used in enforcing good operating practice in Amateur Radio.

When *QST* resumed after the war, one of its first contributors was T.O.M. In an early 1919 issue he contributed an article "Rotten Starting" to work off steam on the slowness with which our government was getting around to let us operate again. At the conclusion of the article appeared the following: "In the Meantime...I am sending you a specimen of a real live Wouff Hong which came to light out here when we started to get our junk out of cold storage. Keep it in the Editorial sanctum where you can lay hands on it quickly in an emergency. We will be allowed to transmit soon and then you will need it."

The object was duly received at HQ, and there it remains to this day.

We know the significance of the Wouff Hong. We don't know the significance of its weird shape. Not even the beloved T.O.M. (revealed, after his death, as none other than our first president, Hiram Percy Maxim) ever explained that. But as the years passed, it continued to grow in the affections of amateurs the country over, old-timer and youngster alike. Today, it is thoroughly entrenched in the lore of Amateur Radio as its most sacred symbol.

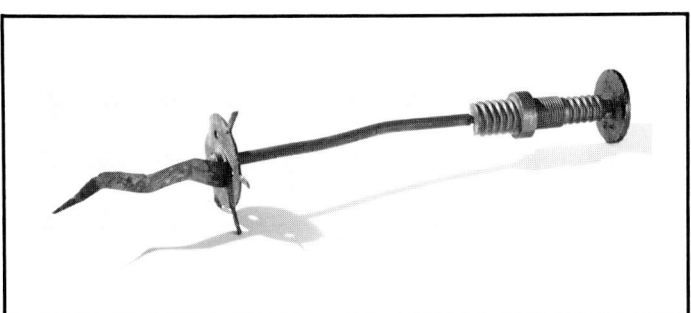

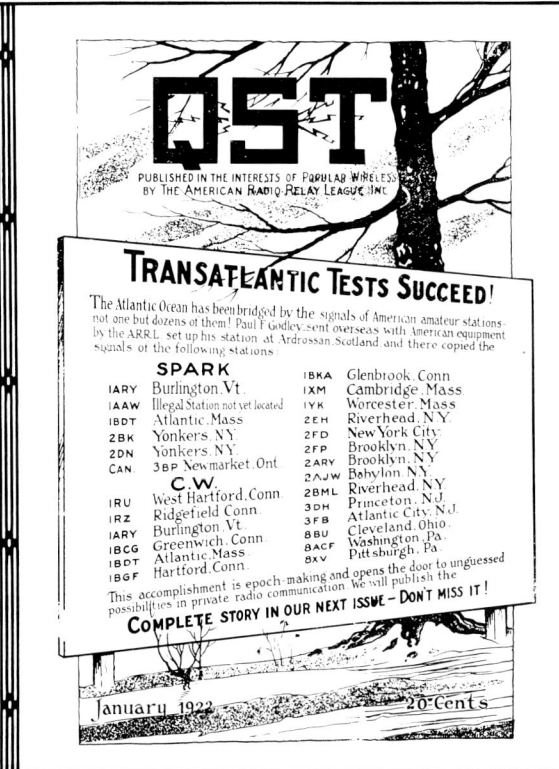

Hartford, Conn., October 6th, 1917

To our readers:

It gives us the biggest regret of our life to have to send out this notice that "QST" will have to be temporarily discontinued. There seems to be no alternative, however, and we must do what we can to make the best of it.

The cause of the discontinuance is that the Editor is going into the Service, and he cannot find anyone who will put up the money and also the work. The Editor has had to put up both and they are necessary. It is a fact that since we were closed up, it has taken more money to run "QST" than the wireless amateurs of the country will supply. "Yours truly" has always had confidence that all he put in would some day come back, but he cannot find any one else who is willing to take the financial chances and do the necessary work in addition.

Therefore, we discontinue and proceed to take our crack at the Germans. As soon as amateur wireless braces up enough to support a magazine, the Directors of the A. R. R. L. will see to it that "QST" is republished. In the meantime, all unexpired subscriptions will be carried forward on the books.

73 until we meet again.

EDITOR OF QST.

Upper left:

W9DAX "My first wireless set —1913."

Center photo:

The "Reti Snitch," a formidable substitute when the Wouff Hong was engaged elsewhere.

June 1929—If you develop a glass arm in sending, says code teaching expert Walter Candler, it could be caused by bodily infection, or by simple muscle tension. For the latter, proper physical exercise is important.

November 1929—Our beloved Hiram Percy Maxim was surprised on his 60th birthday with more than 700 messages of greetings from amateurs and organizations around the world.

December 1929—K. B. Warner reports that after the C.C.I.R. Hague meeting, European nations agreed regionally to limit amateurs to 50 watts, and only 3500-3600 kc. at 80 meters, but the U.S was successful in preventing these limitations from applying worldwide.

December 1929—If the 199 or 222 tube in your receiver is too microphonic, W7NJ suggests switching to a 227.

HILET POWER TRANSFORMERS

Mounted 700 watts, 1000-1500 volts each side, $14.50, unmounted 250 watts, 500-750-1000 each side, $9.75. 100 watt 325 volts each side, two 7½ V windings, $6.50. 100 watt filament any voltage $4.50. Chokes with adjustable core 250 MA $7.50. 160 MA $6.00. Specials to order. Write for lists and specifications. High-grade material and all guaranteed.

RADIO PARTS SALES CO. ORANGE, N. J.

SPECIAL TO AMATEURS

FREE RADIO GUIDE — SEND FOR IT!

Barawik's new short wave dept. has everything that amateurs desire. The Barawik Radio Guide gives full details. Send for it.

Shows the latest wrinkles, newest developments in radio at startlingly low prices. Get the set you want here and save up to 50%. The best in parts, kits, complete factory-built sets and supplies. Orders filled same day received. Write for free 264-page copy NOW. Wholesale prices to set builders, dealers, agents.

BARAWIK CO. 119 Canal Sta., CHICAGO, U. S. A.

The NEW Easy-Working VIBROPLEX No. 6

Reg. Trade Marks: Vibroplex; Bug; Lightning Bug

In Attractive Colors Blue Green Red

Hundreds of operators have traded in old models for this NEW Vibroplex, because it is EASIER to handle. Your old Vibroplex accepted as part payment

Blue, Green, Red or Black ... $17   Nickel-Plated $19

Famous Improved VIBROPLEX

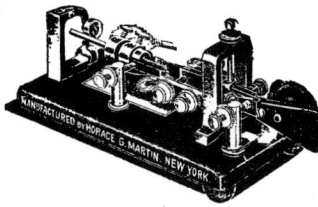

Used by tens of thousands of operators because of its ease and perfection of sending. Colors: Blue, Green, Red or Black $17 Nickel-Plated $19

Special Radio Model Extra Large, Specially Constructed Contact Points for direct use without relay. Colors Blue, Green, Red or Black $25

Specify color when ordering

Remit by Money Order or registered mail

THE VIBROPLEX COMPANY, Inc.
825 Broadway, New York City
Cable Address: "VIBROPLEX" New York

Use Any Tube

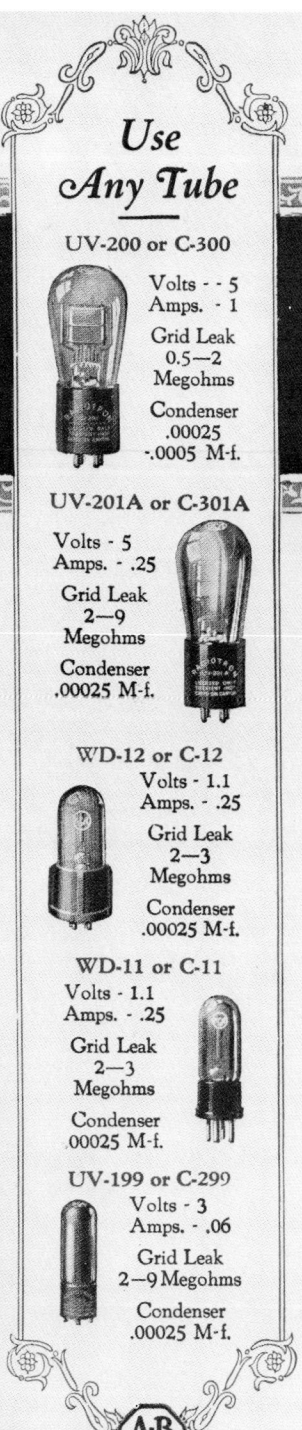

UV-200 or C-300
Volts - - 5
Amps. - 1
Grid Leak 0.5—2 Megohms
Condenser .00025 -.0005 M-f.

UV-201A or C-301A
Volts - 5
Amps. - .25
Grid Leak 2—9 Megohms
Condenser .00025 M-f.

WD-12 or C-12
Volts - 1.1
Amps. - .25
Grid Leak 2—3 Megohms
Condenser .00025 M-f.

WD-11 or C-11
Volts - 1.1
Amps. - .25
Grid Leak 2—3 Megohms
Condenser .00025 M-f.

UV-199 or C-299
Volts - 3
Amps. - .06
Grid Leak 2—9 Megohms
Condenser .00025 M-f.

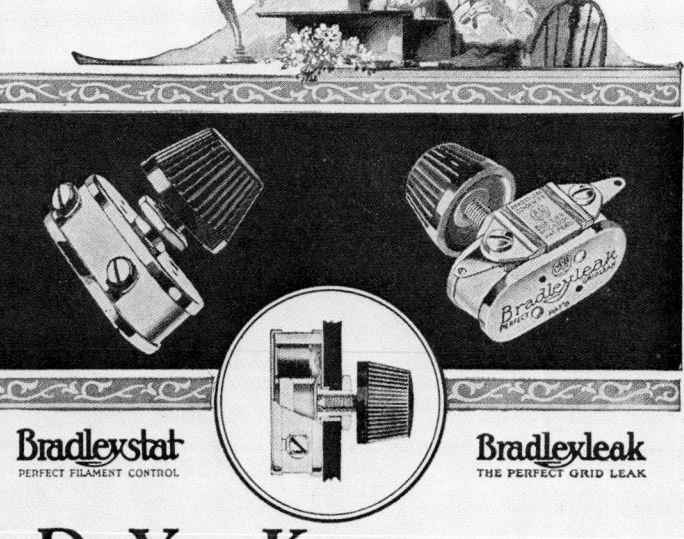

Bradleystat PERFECT FILAMENT CONTROL

Bradleyleak THE PERFECT GRID LEAK

Do You Know—

that any tube can be used in your set without changing rheostats or grid leaks?

IT sounds unbelievable, but it's true. The perplexing problem of selecting the correct rheostat or grid leak is solved by using the Bradleystat and the Bradleyleak. They offer the most marvelous range without steps or noise, and such smooth precision of control that no other rheostat or grid leak can approach them in performance.

The Bradleystat has a resistance range from approximately 1/4 to 100 ohms, by merely turning the adjusting knob that varies the pressure on the graphite discs. It will handle *all tubes* without change of connections, and provide ample control in every case.

The Bradleyleak, with a range from 1/4 to 10 megohms, can be adjusted instantly for any tube, indicated in the adjoining table of tube ratings, by turning the adjusting knob.

Be ready to use any tube in your radio set. Install Allen-Bradley Radio Devices, throughout.

Allen-Bradley Co.
ELECTRIC CONTROLLING APPARATUS

Sales Offices: Baltimore, Birmingham, Boston, Buffalo, Chicago, Cincinnati, Cleveland, Denver, Detroit, Knoxville, Los Angeles, New York, Philadelphia, Pittsburgh, Saint Louis, Saint Paul, San Francisco, Seattle

General Offices and Factory: 277 Greenfield Ave. Milwaukee, Wis.

Manufacturers of graphite disc rheostats for over 20 years

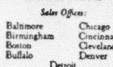 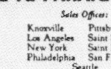

Have you used the Bradleyswitch? It saves batteries and tubes.

FERRANTI
AUDIO FREQUENCY TRANSFORMERS
FOR ANY CIRCUIT

Ferranti Transformers stand supreme in that they pass on to the loud speaker an uncensored message. All the notes...bass notes and treble notes... are there.

FERRANTI, Inc.
130 West 42nd St., New York, N.Y.
FERRANTI, LTD.   FERRANTI ELECT., LTD
Hollinwood, England   26 Noble St., Toronto, Canada

Bradley-Amplifier
Resistance-Coupled
PERFECT AUDIO AMPLIFIER

Allen-Bradley Co.
Electric Controlling Apparatus
277 Greenfield Avenue   Milwaukee, Wis.

BECOME A RADIO OPERATOR
See the World, Earn a Good Income, Duties Light and Fascinating
LEARN IN THE SECOND PORT U.S.A.

Radio Inspector located here. New Orleans supplies operators for the various Gulf ports. Most logical location in the U.S.A. to come to for training.

Nearly 100% of radio operators graduating on the Gulf during the past seven years trained by Mr. Clemmons, Supervisor of Instruction.

All graduates placed. Runs to all parts of the world.
Member of the A.R.R.L. — Call "W5GR."
Day and Night Classes — Enroll any time.

Write for circular

GULF RADIO SCHOOL
844 Howard Ave.   New Orleans, La.

FOR THE NEW UX866 RECTIFIER TUBES THE T3680 FILAMENT SUPPLY TRANSFORMER IS AN IDEAL UNIT

Just wait 'till I shoot my sigs out in the air with my new Thordarson Power Supply. I'll make 'em all sit up an' take notice.

THORDARSON
ELECTRIC MFG. COMPANY
500 WEST HURON STREET
CORNER KINGSBURY
CHICAGO, ILL.

THE '30s

Upper left:

1931—George Grammer, Technical Editor of *QST*, and James J. Lamb, Research Engineer, labor in the ARRL laboratory.

Upper right:

W1AWW relayed signals on 5 meters way back in 1932 from this 90-foot tower on Wilbraham Mountain in Springfield, Massachusetts.

Center photo:

Power supply at station W8CPC. Left: motor generator, 2200 volts—can vary from 700 to 2200 volts. Right: Mercury arc, 2500 volts. Switch board (left to right) RF chokes, high-power switches, 250,000-ohm bleeder, ac switch. Note that the base of the motor generator support is resting on Ford clutch springs. No vibrations and a smooth ride, Hi.

Lower left:

W9SUJ keeps an eye on the line voltage from his belt-driven alternator, run continuously for 27 hours.

Lower right:

Ray E. Meyers, Radio Operator on Sir Hubert Wilkins' submarine. The 200-watt radio transmitter was built by General Electric in Schenectady, New York, for Radio-Marine Corp.

January 1938—The I.A.R.U. News column presents a new list of "countries" for DXCC purposes, revised from a year ago to take into account many suggestions from overseas as well as domestically. G6WY now tops the award list with 115 confirmed.

June 1938—More than 1,000 participated in the 1938 Field Day, an event challenging the DX and Sweepstakes Contests for total interest. The Egyptian Radio Club's (Illinois/Missouri) W9AIU/9 established an all-time high with 317 contacts and a score of 3708, operating on four bands simultaneously.

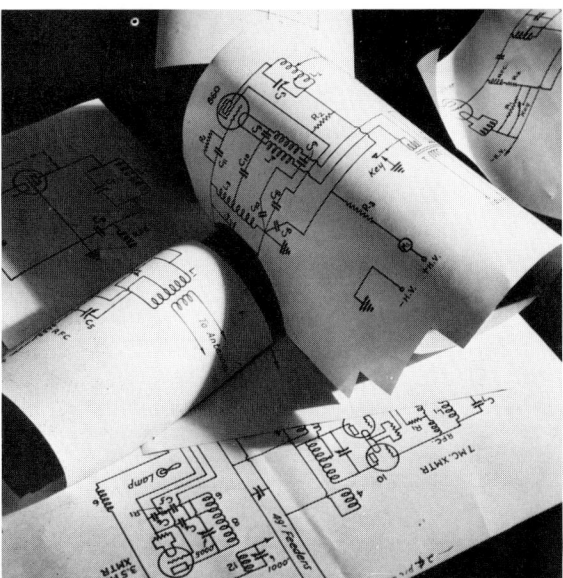

Upper left:

Building a crystal-controlled transmitter
Beginners—and a large number of amateurs who were not beginners—frequently were apprehensive about the use of crystal control because it looked complicated and expensive. Many found that not to be the case. This photo shows how to build an inexpensive outfit that any beginner can make.

Upper right:

High adventure with a portable in an auto caravan

Center photo:

A view from the other side showing the plate choke and the 14-mc. tank coil of a typical transmitter of the period that was not crystal controlled. Mechanical stability was important. The tube is a 203-A.

Lower left:

Circuit Diagrams
These are the schematic diagrams for a medium-power HF transmitter. Circuit symbols have evolved over the years, but the old diagrams are still easily read.

Lower right:

A modular constructed receiver from the decade.

Upper left:

The four-band "Kitchen transmitter" in its operating position beneath the gas range. The color combination matches that of the stove, making the outfit "hot" in every sense of the word.

Upper right:

W2DKJ redefines "mobile communications" at the New York World's Fair.

Center photo:

High power stations in the early thirties used a mercury-arc rectifier as a source of "high potential" dc. The two small rectifiers on the right supplied voltage to keep the arc alive. To start the arc, the tube in the center was tipped, and mercury covered the arc. When the tube was tipped back to an upright position, the arc was struck and operation of the rectifier began. By the mid-thirties, higher power mercury-vapor rectifiers were developed to replace the mercury-arc rectifiers.

Lower left:

Experimental model of the Linear Electronic Voltmeter. Input connects to the jack-top binding posts at the upper left, the important resistor R is immediately below. The resistor for the other range plugs into a pair of holes in the baseboard. The 0-200 microammeter connects to the posts at the lower right in normal operation and to the pair of posts immediately above when it is used for checking filament voltage.

Lower right:

Little ON4UU on his daddy's feeders. Brussels, June 1932.

December 1938—A piece of quartz labeled just inside a band edge is no assurance that circuit constants and temperature changes won't cause drift over that edge, says the Editor, who points out F.C.C. monitors can measure as close as ten cycles and are sending "pink tickets" to violators.

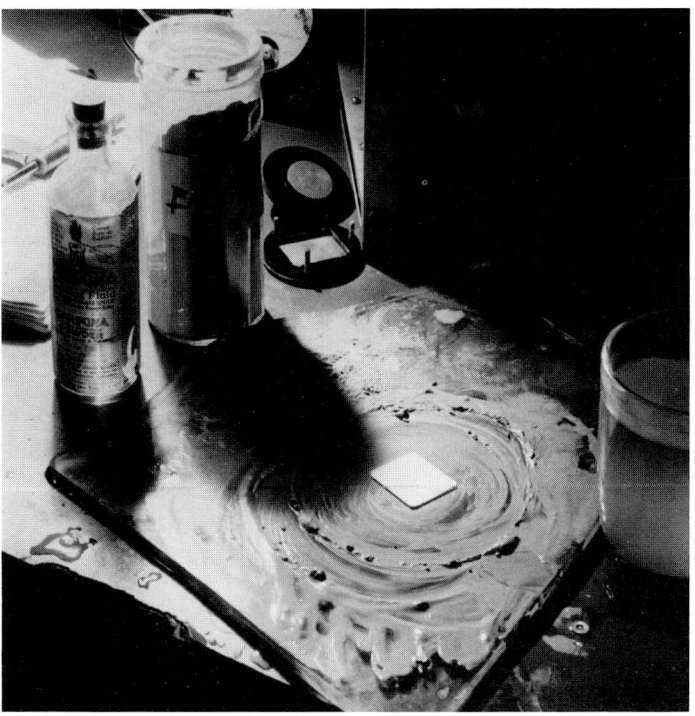

June 1939—Fierce competition in the broadcast receiver industry has led to cutting corners in design and manufacture, and hams are being unfairly blamed for increasing BCI. Editor Warner says the problem is cheap sets, and argues that since they are not of "modern design" as defined in F.C.C. standards, the "quiet hours" regulation should not be invoked.

Upper left:

QST Technical Editor James J. Lamb, showing off his classic noise silencing circuit.

Upper right:

No phase noise here! Hams in the thirties ground their own crystals to their favorite frequency.

Lower left:

A general view of a home-made four tube receiver. The tuning knob at the side was provided to make tuning a more comfortable process than it ordinarily was. Below the tuning knob is the volume control disk, mounted horizontally and projecting through a slot in the shielding.

Lower right:

Traveling light? Not really. Don Wallace, W6AM has a complete 1-kilowatt rig packed in those two bags! *(W6DEP photo)*

June 1939—A committee to set up programs to preserve amateur bands, a study of Hdq. operations, and a poll to determine amateur sentiment on opening 7200-7300 kc. for voice, are among actions of the Board of Directors, meeting in San Francisco this year.

All You Need to Learn TELEGRAPHY
Morse or Continental with Teleplex

Learn Telegraphy — the most fascinating profession — by hearing *real* messages — *sending* them. Interesting — simple — you learn quickly — at home.

TELEPLEX — the *Master Teacher* is used by U. S. Army, Navy and leading radio and telegraph schools. Entirely new code course in 12 rolls of tape.

During last ten years, TELEPLEX has trained more operators than all other methods combined.

Write for Folder Q-5

TELEPLEX COMPANY
76 Cortlandt Street   New York

Confidence in a Good Name

You get advanced engineering in Ken-Rad Tubes. For immediate good results specify Ken-Rad Tubes. All types of tubes available.

Ken-Rad Radio Tubes
KEN-RAD TUBE & LAMP CORPORATION - OWENSBORO, KY.
Makers of all types of radio tubes and Ken-Rad Electric Lamp Bulbs

Say You Saw It in *QST* — It Identifies You and Helps *QST*

Answers YOUR Questions
HOW TO PASS Radio License Examinations

By CHARLES E. DREW, I.R.E., A.I.E.E.

This book is more than just a new edition of a famous work. It has been completely rewritten, and will serve as an excellent guide to all radiomen. Important phases of the subject are given thorough treatment, so that the man who understands the content of this book will be well qualified to pass the examination for his commercial license. **$2.00**

ON APPROVAL COUPON

John Wiley & Sons, Inc.
440 Fourth Avenue, New York, N. Y.

Kindly send me a copy of Drew's HOW TO PASS, on ten days' approval. If I decide to keep it, I will remit $2.00; otherwise I will return the book postpaid.

Name.......................................
Address....................................
City and State.............................
Employed by...............................

QST 4-39

The HAM SUPER
MAKES ZEDDERS AND AUSSIES R7-R8 IN EARLY EVENING

Already the new Ham Super is making records for other ham receivers to shoot at — and it was introduced only last month. Read a bit of what 9CVE, George Miller of Chicago says of it after two weeks' operation:

"The new Ham Super is undoubtedly the best receiver for sensitivity, selectivity and single signal tuning that I ever heard of used. Its performance is remarkable on all Amateur frequencies, band-spread or coverage, CW or fone, and it is a real pleasure to carry on communication with such reliability, without the usual QRM and heavy background interference so common to most receivers today.

The sensitivity is so high that I bring in New Zealand and Australian stations much earlier in the evening and with more signal strength. Some of their R7-8 signals could easily be mistaken for stations on this continent before they give their call letters. Another surprise was to hear west coast stations fairly consistently over a period of two weeks up until 10:30 A.M. (C.S.T.) and as early as 3 P.M. (C.S.T.) on 7 megacycles — rather unusual reception for this frequency band.

That novel idea of changing frequencies is such an easy operation and works great.

I am thoroughly satisfied. It does all that you said it would."

If you want a really advanced, up-to-the-minute Ham receiver, find out about the Ham Super. Good? — it's got to be good to do that, and it was designed by amateur authorities for whom you have the greatest respect. Send in a stamp for complete details of the Ham Super, the new electron coupled frequency meter-monitor (all A.C.) and other hot ham transmitting and receiving specialties.

LAST MINUTE SPECIAL
A new and compact all A.C. electron coupled frequency meter and monitor combined. It covers all Ham bands and is only 6″ square. And the price? — depression low.

McMURDO SILVER, INC.
1136 W. Austin Ave.   Chicago, Ill.

McMURDO SILVER INC.
1136 W. Austin Ave., Chicago, Ill., U.S.A.

Enclosed find 3c stamp for which send me details of the Ham Super and all amateur apparatus.

Name.......................................
Address....................................
Town.................................State..........

YOUR FIRST LINE OF DEFENCE AGAINST SUMMER SLUMP

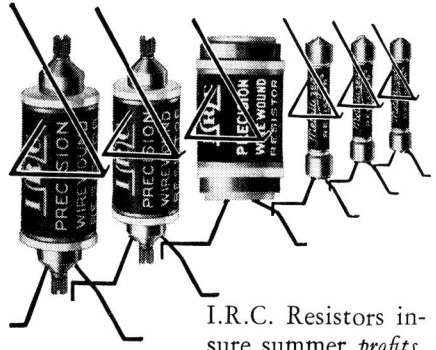

I.R.C. Resistors insure summer *profits*. Metallized Resistors for replacements. Wire Wound Resistors for meters and test equipment. Make for yourself valuable apparatus which will speed up your service work, build your reputation and add satisfied customers.

Mail coupon today for FREE charts. They will save you hundreds of dollars in equipment.

INTERNATIONAL RESISTANCE COMPANY
Philadelphia   Toronto

Mayo Type 'Q' Microphones

Formerly offered only to broadcasters, recording studios, etc.

This is truly an instrument you will be proud to own. It uses the new ground center, heat treated, duraluminum diaphragm which insures sensitivity, absence of hiss, and a frequency response equal to microphones listing up to $75.00. This new and improved microphone is a precision instrument built to rigid specifications and is broadcast size, measuring $3\frac{1}{8}''$ diameter x $2''$ thick, 100 or 200 Ohms per button and finished in pure silver.

Guaranteed

MAYO INSTRUMENT CORPORATION
1 Madison Avenue   New York City

$9.50 NET To Amateurs

METAL-WORKING LATHE Precision equipment for the small shop. New designs and manufacturing processes make this amazing bargain possible. Complete metal working lathe with compound slide-rest, combination face-plate and independent chuck and tail center, 6'' swing; 24'' length, 20 pounds. Send $1.00, balance plus postage C.O.D. Lathe for wood-turning alone, $4. Attachments for milling, grinding, sanding, saw-table, etc., available at low prices. Order from ad at once and have a complete machine shop.

$7.50 COMPLETE

AMERICAN MACHINE & TOOL CO., Dept. Q-8 200 BROADWAY NEW YORK

ALUMINUM BOX SHIELDS Genuine "ALCOA" stock, silverdip finish, 5 x 9 x 6, $1.65. 10 x 6 x 7, $2.65. *Any Size to Order.* **SOMETHING NEW!** Your call letters, or any marking for your panel, on BLACK aluminum ribbon. Looks like engraving on bakelite. 5c up to 2 inches, 5c each additional inch. Sample 8c. U.S. Army V.T. 1 tubes 35c, 10 for $2.50. Foil for condenser or velocity mike ½ mil., 25c ft. New Master Teleplex on demonstration.

BLAN, the Radio Man, Inc. ·W2GT· 177 Greenwich St. New York City

49

THE '40s — THE WARTIME YEARS

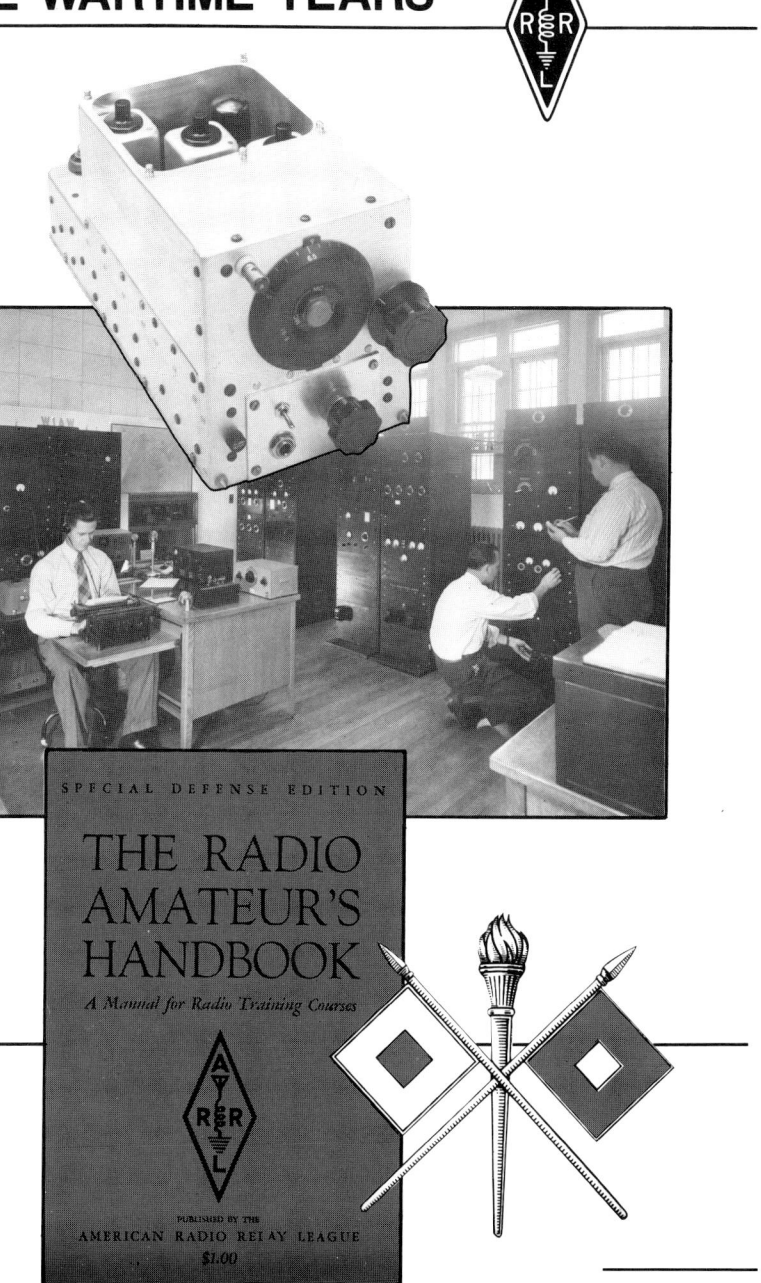

Upper left:

Summer 1948—W4PL, W1RUP, W1QVF admire good taste in license plates!

Upper right:

MRI XMAS HA...

Lower left:

One of the monitoring bays at the FCC station at Grand Island, Nebraska. The antenna-switching panel is in the right-central rack, just above the row of Keytype switches at the bottom. Note the Hallicrafters SX-28s and Hammarlund Super Pro.

Lower right:

Cpl. James Ralph Graham, who attained a speed in transcribing International Morse code on a typewriter that was said by fellow instructors in Central Signal Corps School at Camp Crowder, Missouri, to be faster than the world's record. Graham, also an expert teletype operator, is shown working at a M-19 teletype machine. *(US Army Signal Corps photo)*

July 1943—Charles Service, W4IE, urges amateurs to be on the lookout for spies and saboteurs who are known to be loose in the country.

December 1940—The FCC gets way behind in the issuing of licenses, principally because of the requirement regarding proof of citizenship. Many applicants for licenses have been enclosing applications in the same envelope with the proof of citizenship documents whereupon the applications become lost in a different department!

Top photo:

Information center on Bougainville for all air activities and anti-aircraft control. The "think before you transmit" sign was typical of security signs throughout the services. It means to observe radio silence (whenever that's imposed), to observe proper procedure on the air, and to not say anything that might be useful to the enemy.

Lower left:

Remember the thrill of your very first QSO? Remember the kick you got from your first real DX contact? And, if you have photography as an allied hobby, remember the excitement you felt when you finished that first print and the picture details began to appear? Put them all together and you have some idea of what was in store when the new ham television camera was fired up for the first time— bring up the Ike bias a hair, a final touch to the focus—and there's the first image beginning to take form on the monitor!

Lower right:

Inside a Martin Mariner—Interior view of radio and pilot compartments in a PBM Mariner, twin-engine Navy patrol bomber. In the foreground (left) is the navigator; the radio operator is on the right. Radio operators were needed on all bombers and transports. *(Official US Navy photo)*

Upper left:

Parachute instructors in Michigan wanted some means of communicating with student jumpers in the air, so they developed a tiny transceiver unit. The acorn-type detector-oscillator tube and the two miniature audio tubes operate from a pair of midget-sized 33-volt "B" batteries and a single flashlight cell. The para-talkie is given a test in the field by Lt. Ralph Berkhausen, CAP, using the special parachutist's antenna attached to his leg.

Upper right:

Production line workers at the Hallicrafters plant assembling HT-4 transmitters (also known by their US Army designation as BC-610). This transmitter was used by a number of hams after WWII (CW, AM phone and RTTY at 500 W—hams could, of course, get more out of them). Seen here are the backs of the transmitters—power supply deck bottom; modulator deck middle. Behind the black screen on top was the RF deck, which used plug-in exciters and coils. It took four men to lift the finished unit.

Lower left:

A home-brew transmitter with high-voltage parts exposed for all to see, with a Collins 75A-1 receiver.

Lower right:

How to tune a 5-meter dipole.

June 1945—The A.T. & T. Co. has filed application for the construction of seven microwave relay stations between New York and Boston, the first of their kind.

Top photo:

At this well equipped station you'll find two National Company HRO-7 receivers, a Collins 32V-1 transmitter, Collins 30K-1 transmitter, Collins PTO, Astatic T-3 microphone, Vibroplex semi-automatic key (bug). *(W6GHI photo)*

Lower left:

The versatile Harvey-Wells Bandmaster transmitter covered 160 through 2 meters. The two receivers are a Collins 75A-1 and a National NC 57, the inexpensive National beginner's receiver.

Lower right:

Sergeant Blackwell has an assistant in his wallaby "Oscar," who not only likes to use the radio, but loves to chew gum. Blackwell is using the "bug" in sending from this underground radio shack in Australia.

November 1945—We are back on the air, but only on u.h.f. The other bands are still held and being used by the military and it may take some time for their release. FCC is not in a position to accept applications for new licenses due to lack of funds. Congress has failed to provide more money. W1AW is on the air nightly by special authority for the purpose of transmitting up-to-the-minute progress on 80, 40 and 20 meters.

December 1949—Five thousand participants made our '49 Field Day the biggest ever.

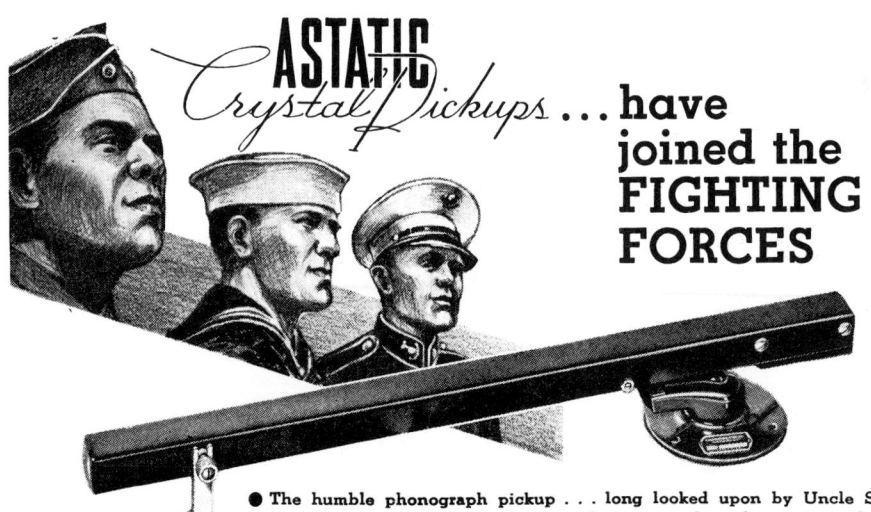

ASTATIC Crystal Pickups ... have joined the FIGHTING FORCES

● The humble phonograph pickup ... long looked upon by Uncle Sam as a luxury product subject to limited manufacture ... has of recent months become an important factor in the educational and morale building program for our armed forces. The Astatic Corporation today supplies Astatic Crystal Pickups to the Special Service Division of the War Department, to the Navy Bureau of Personnel, to the Marines and other branches of the service employing phonographs in recreational centers, on ships, landing craft, and other places where instructive and entertaining recordings are broadcast. Along with Microphones and other Astatic products, Crystal Pickups are available with proper priority ratings.

In Canada: **Canadian Astatic, Ltd., Toronto, Ont.**

THE ASTATIC CORPORATION
YOUNGSTOWN, OHIO

SALUTE TO THE SIGNAL CORPS

LEO-W9GFQ JOINS HALLICRAFTERS

In Saluting

THE SIGNAL CORPS

I've always been a big distributor of Hallicrafters equipment. I'll pay you highest prices for your used Hallicrafters rig and your other sets and parts, too. Send description and I'll pay you cash immediately without any bother or inconvenience on your part.

WRITE TODAY FOR FREE FLYER LISTING HUNDREDS OF HARD-TO-GET ITEMS!

WHOLESALE RADIO LABORATORIES
744 West Broadway
Council Bluffs, Iowa

SPECIALIZING IN HALLICRAFTERS EQUIPMENT

GENERAL COMMUNICATIONS CRYSTALS HOLDERS AND OVENS
Precision Made by **Bliley**
WRITE FOR CATALOG G-11
BLILEY ELECTRIC CO., ERIE, PA.

TELEGRAPH SPEED KEYS
Radio Type in Kits — $2.89
Send Card for Complete Information
ELECTRIC SPECIALTY MFG. COMPANY
Box 645, Cedar Rapids, Iowa

RADIO COURSES
Classes Start October
● RADIO OPERATING ● BROADCASTING ● CODE
● RADIO SERVICING ● TELEVISION
● ELECTRONICS — 1 year day course; 2 years eve.
Day and Evening Classes — *Booklet upon request*
NEW YORK YMCA SCHOOLS
4 West 63rd Street, New York City

PRE-MILITARY TRAINING
for MEN of MILITARY AGE
CIVILIAN TRAINING
for MEN and WOMEN
seeking Careers in Radio
Complete Course up to 8 Months
Write, Phone or Call 9 a.m.- 9:30 p.m.

METROPOLITAN TECHNICAL SCHOOL
RADIO DIVISION, Dept. S
7 CENTRAL PARK WEST, N.Y.
Circle 7-2515   Licensed by State of N.Y.

THANKS FOR THE PLUG, ROMMEL

...early in 1941 I was transferred to Tunis where I have remained until the present. During all these times your receiver gave me the best of service and enabled me to follow broadcasts from the United States as well as Europe.

Unfortunately, during the German occupation of Tunisia after our landing in North Africa, my house in Tunis was occupied by the German Commander-in-Chief who apparently found your receiver as much to his liking as I had. In any event, upon my return to Tunis after the recapture of that city, I found it missing together with the greater part of my furniture and household effects.

It would be appreciated if you would again provide me with the present equivalent of the set which I possessed.

(Excerpt from a letter we received from a member of the State Department)

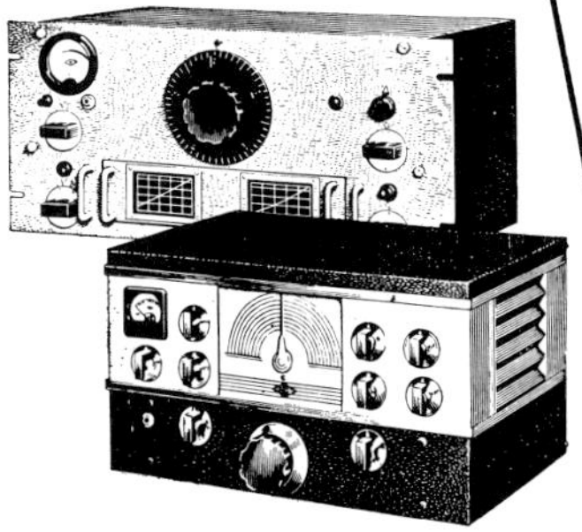

NATIONAL COMPANY, INC.
MALDEN, MASS, U. S. A.

★ **DO YOUR BIT** ★
TO HELP A COMPANY DOING 100% WAR WORK

We need 150- and 300-volt scale voltmeters, 3" to 5" face, about 1% accuracy. New or used. Will pay ceiling price. **Kato Engineering, Mankato, Minnesota.**

WE "CARRY ON"

UNIVERSAL has consistently manufactured microphones for the Signal Corps and is currently producing hand, throat and lip styles for the Armed Forces. UNIVERSAL will continue to do its part in making our soldiers "the best equipped in the world."

UNIVERSAL MICROPHONE COMPANY
Inglewood, California

THE '50s

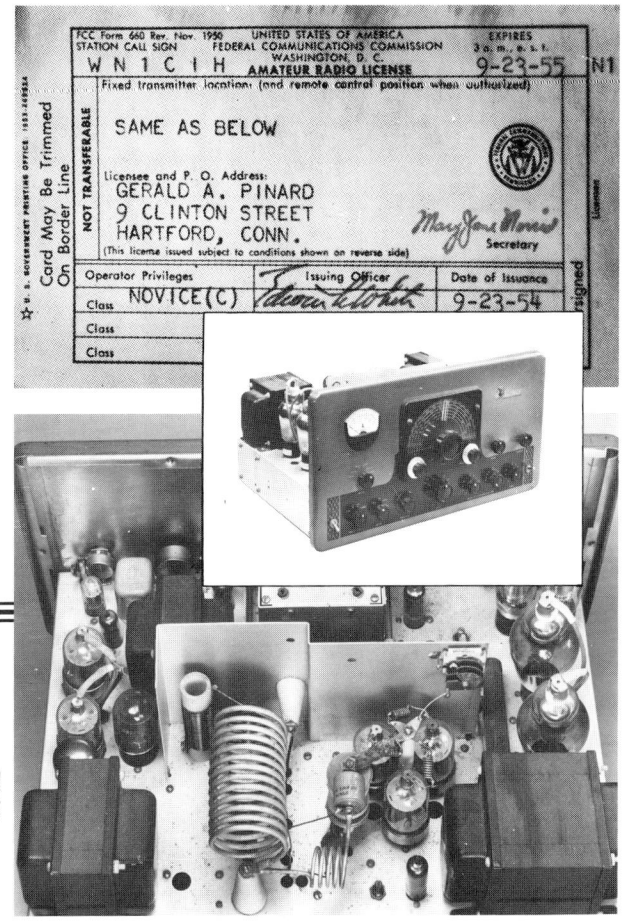

**A NEW BAND EFFECTIVE
MAY 1, 1952**

Upper left:

This neat little package combined Don Mix's "Bandbox" frequency-multiplying unit with a 6146 amplifier using a continuously variable inductor in a pi-network tank. The construction is such that the unit is self-shielding for TVI—with only one very simple metal piece requiring fabrication.

Upper right:

Grinding out QSOs from W4GAC/4!

Lower left:

Passing of a landmark. On June 19, 1954 the Wilbraham Tower (details in the 30s section) was destroyed by a fire of unknown origin.

Lower right:

Here is something that whet the imagination of the Buck Rogers and Dick Tracy sets: radio communication with the necessary electrical power derived directly from sunlight. The solar battery can be seen clamped to the top left of the operating table. The transistor transmitter is in the center with the receiver to the left.

December 1950—"Copyability" is more important than fidelity in ham communication, and this is the theme of W1DX's 50-kc. i.f. amplifier; he warns that signals will "tune sharp," achieved by cascading several tuned circuits in each tube stage. The brothers W3NJE and W8KML tackle the transmitting side, with a speech amplifier and driver design using clipping and a low-pass filter.

Field Day — 1953

December 1950—Recent issues of *Radio News* have touted "supermodulation" as the ideal means of ham voice communication; Prof. Villard, W6QYT, presents the results of his extensive evaluation, showing many of the claims are mutually conflicting, and concluding there is no magic or revolutionary technique in this system.

December 1951—The Korean war has restricted parts availability, but hams are granted self-rated priorities for building and replacement; those enrolled in defense and security activities such as AREC, NTS, MARS get a double quota.

Upper left:

W9ERU/9 Field Day—1957. *(K9BJA photo)*

Upper right:

Normally pretty cheerful guys anyway, General Manager A. L. Budlong (left) and Circulation Manager David H. Houghton (right) had a good reason for those wide grins—the millionth copy of ARRL's *License Manual*.

Lower left:

Nine YLs who enjoyed a variety of activities at the Eastern Canada ARRL convention in Montreal, September 19, 1953, are (left to right) VE2RK, Therese; VE2AOB, Stella; VE2CA, Phyllis; VE2NJ, Nancy; W1ZCS, Marie of ARRL Hq; K2DRY, Emily; K2CBS, Ida; VE2AKK, Betty; and VE2HI, Ethel. *(R.E. Fleischman photo)*

Lower right:

Gwen Rudolph, KN6IHD, and her Gonset Communicator II. The Gonset was one of the few rigs that would operate on 2 meters. It had only one eye—in the middle—which closed when you received more signal. AM, of course.

December 1950—Keeping on frequency will be easier with the 100-kc. oscillator with 10-kc. multivibrator described by ZL3LR.

Upper left:

Teleprinter art, popular in the 50s, displays a festive message.

Upper right:

W5BHO takes a break with a good book. His station features a Hallicrafters SX-25, a little homebrew, and surplus command sets. The end result provides access to the 2-6-10 and 75 meter bands

Lower left:

Amateur Radio enthusiasts come younger all the time. This young gent was exposed to 75-meter QRM at the innocent age of several hours. Mom is Mildred Drummond, W0GXG.

Lower right:

D. T. Baird, RMC, USNR (W5SPZ) and Cmdr. W. R. Sherman, USNR, District Reserve Electronics Program Officer, manning amateur station W5USN/5 of the District Reserve Master control station at the 1953 ARRL Delta Division Convention —New Orleans, Louisiana.

December 1953—The new double-conversion strips to get u.h.f. reception on older TV receivers pose a threat to amateurs in that the first i.f. falls near the 2-meter band, W1HDQ warns.

June 1954—With FCC no longer conducting Novice and Technician exams, it now becomes our responsibility to maintain high standards of integrity, as the Editor pointedly reminds us.

December 1951—To assist licensees in the new Novice Class to upgrade operating skills, ARRL announces the "Novice Roundup," with an invitation to OT's to take part and give the newcomers contacts.

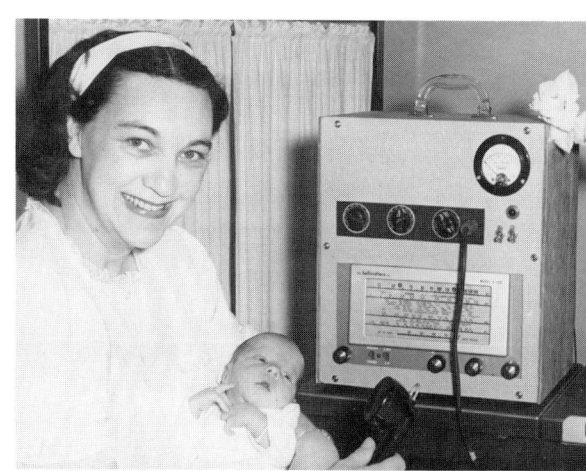

November 1954—Sooner or later a c.w. fan wants to try voice, and W1JEQ has designed a 25-watt modulator you can use when the bug bites.

December 1954—Editor Warner explains that the League support of FCC's proposal to put Technicians on 50 Mc. is primarily based on the need for occupancy, and our opposition to their use of 144 is simply that it would dilute the basic purpose.

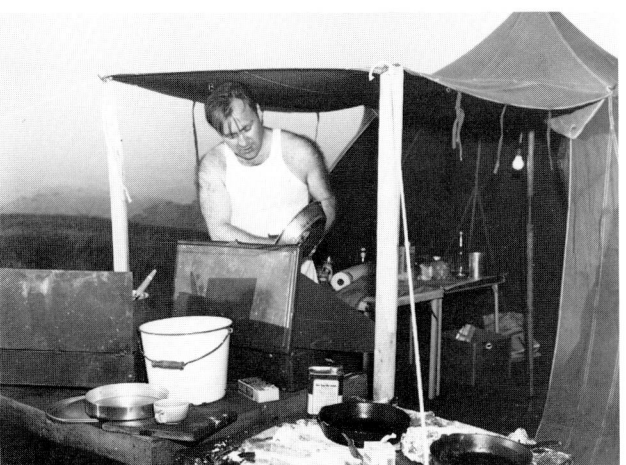

Upper left:

This antenna system could have been the means of achieving the long-sought goal of 144-Mc. DX up the Pacific Coast. A 30 foot parabola mounted on a dolly, so that it can be rolled around on the flat roof, it is erected on a 1200 foot elevation directly above Hollywood. The lights of the Los Angeles area stretch out for 20 miles toward Long Beach in this night shot by KN6GLG. K6EGP is seated at the left, W6COI climbs the framework on the rear of the reflector, and K6BXW is at the right.

Upper right:

W8PQQ's shack showcases gear from the 30s, 40s and 50s.

Center photo:

While her mother, W7ULK looks on, WN7VWU operates this classic mid-fifties station. To the left is a top-of-the-line Hallicrafters SX-88 receiver, and the transmitter is a popular Johnson Viking II. The Viking VFO (behind the key) was not used in the Novice bands. On the top shelf shelf is a Pennwood Numechron 24-hour clock and 275-watt Johnson Matchbox. Old-timers will recognize some of the QSLs as being the Orange and Blue design sold by the Walter Ashe Radio Company of St. Louis.

Lower left:

Bill Koutnik, W6ZXH, unconcernedly toiled away at KP while his cohorts had all the fun at the Aerojet Radio Amateurs Club FD site, Carbon Canyon, California.

W1BFT CALLING ALL HAMS

offers you! *Tecraft*

Model CC5 for VHF use up to 220 mcs, Complete—$42.50, specify I.F.; Kit form $29.95.

Evans RADIO
P.O. BOX 312 • CONCORD, N.H.

LEARN CODE ...FAST!

Build up your code speed quickly...easily. Just a few minutes a day spent with this practice set can boost you over that hump. Set includes a constant frequency buzzer and key mounted on a 4"x6" molded Bakelite base. May be used singly or in pairs for code practice.

Cat. No. 114-450   **$4.25** Net Price

E. F. JOHNSON COMPANY
WASECA, MINNESOTA

What's goin' on? — WALTER ASHE

EVERYBODY'S BUYIN' MOBILE EQUIPMENT AT WALTER ASHE WHERE THEY GET THE "SURPRISE" TRADE-IN ALLOWANCE ON USED FACTORY-BUILT EQUIPMENT

Whether you are a beginner or an "ole timer", here are four big reasons why you are way ahead with a Walter Ashe "Surprise" Trade-In:

1. **HIGH ALLOWANCE** on your old factory-built gear
2. **IMMEDIATE DELIVERY** in factory-sealed cartons
3. **CLEAN CUT DEAL**—you get our BEST price FIRST
4. **PROMPT REPLY**—we don't keep you on the hook

All prices F. O. B. St. Louis • Phone CHestnut 1125

Get your trade-in deal working today.
Wire, write, phone or use the handy coupon.

```
Walter Ashe Radio Co.                              Q-5-52
1125 Pine St., St. Louis 1, Mo.
Please make "Surprise" Trade-In offer on my used (factory-built)
_____
              (describe used equipment)
for the following new equipment_____
_____
(specify make and model number of new equipment desired)
Rush free copy of your new catalog.
Name_____
Address_____
City_____ Zone____ State_____
```

Walter Ashe RADIO CO.
1125 PINE ST. • ST. LOUIS 1, MO.

Winpower Generator Set Supplies Power For Radio Hams' Call For Help —

As Schooner Goes Adrift 800 Miles At Sea

On April 8, 1954, five Iowa men, Tom Partridge, Vern Hedman, Robert Denniston and Gene O'Leary of Newton, Leo Olney of Des Moines and two Mexicans with a crew of six left Acapulco, Mexico, aboard an 80' schooner bound for the lonely Pacific Island of Clipperton. Their objective was to set up an amateur radio station and contact as many amateur radio operators as possible.

This trip, in the minds of all aboard, was to be routine, however, they did expect some difficulty in landing on the island. But undetermined fate struck. Within sight of the Island, a severe windstorm struck the expedition, ripping the schooner sails and due to a faulty waterpump, engine trouble developed, leaving the schooner without motive power. The party was in serious trouble but communication with the outside world was maintained. Fastened to the deck of the schooner, exposed to the ravages of the sea, salt water, wind and tropical sun, was a Model G-800A WINPOWER Generator Set, which despite the elements and the fact that it had been in almost continuous operation 24 hours a day for three weeks since the party left Mexico, produced unfaulty electric current for operating the party's radio transmitter receiver and for short periods the boat's bilge pump. Yes, thanks to a WINPOWER Generator Set, the party did go ashore on Clipperton Island and although the set was temporarily out of service after being dunked in the surf, did complete their objective and talked to over 1000 amateur radio operators.

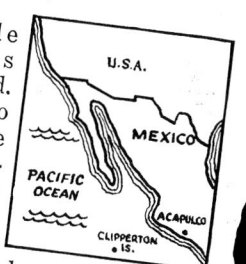

This proves the statement once again that WINPOWER Generator Sets deliver dependable power when it is needed and where it is needed.

28 years of engineering and manufacturing "know-how" goes into the building of all WINPOWER Generator Sets which makes them a dependable source of electric current for anyplace on earth.

WINPOWER Electric Plants are used for portable plants or stationary plants and can be used for constant or intermittent service. There is a size for every need from the small 500-Watt to the large 12,000-Watt Set. Briggs & Stratton and Wisconsin Engines furnish the motive power for this fine equipment.

When you need a Generator Set, buy the set that you can depend on. Buy a WINPOWER.

WINPOWER MANUFACTURING COMPANY
NEWTON, IOWA

Mobile Antennas --- by Premax

Car-top, Whip, Center-Loaded, Zone and Sector Civil Defense Control . . . a complete line for every purpose including Marine. Send for special bulletin of antennas and mountings

PREMAX PRODUCTS
DIVISION CHISHOLM-RYDER CO., INC.

5302 Highland Avenue, Niagara Falls, N. Y.

For Selectivity
Never Before Achieved In a
Communications Receiver

Dr. Qwak

℞ The Collins 75A-3
With Mechanical Filter

Dr. Qwak (Willard Wilson — W3DQ) also has B & W, Collins Xmtrs, National, Hallicrafters, Johnson, Elmac, Gonset, etc. . . . all for prompt delivery, and on the easiest of terms. Write today.

Wilmington Electrical Specialty Co., Inc.
405 Delaware Ave., Wilmington, Delaware
Est. 1920
Willard S. Wilson, President
Member OOTC — VWOA — QCWA
A·A·O·N·M·S

THE '60s

Upper left:

You don't have to be an old-timer to remember when communication receivers were heavy, cumbersome devils that brought on a lot of "teeth-clenching" when they were moved from one place to another. That's why it was such a pleasant surprise to find the 75S-3 receiver weighed only 20 pounds!

Upper right:

These young ladies participated in the 1963 Miss Amateur Radio pageant held at the Michigan State Amateur Radio Convention. They are (from left to right) Kay Mellberg, daughter of W8QPO; Susan Gough, daughter of K8OIC; and Carol Thomas, daughter of W8SZY. Miss Mellberg was the lucky winner. The lucky guy at the left is Don McMillan, K8KWG, representing the Saginaw Valley Amateur Radio Association.

Lower left:

1960s style auto caravan.

Lower right:

Central-Division single-op VHF contest champ K9KFR fondly regards his HS-type cat and perhaps ruefully considers the QSOs he missed on 2 because the kilowatt amplifier expired.

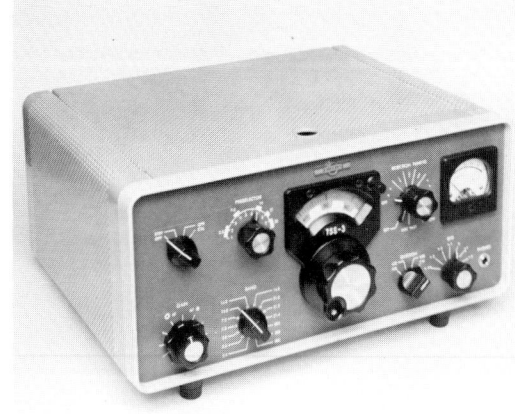

December 1963—The international conference on Space Communications in Geneva, with IARU admitted as a participating organization in observer status, has approved space satellite activity in the 144-146-Mc. band, thus "legalizing" the Oscar programs.

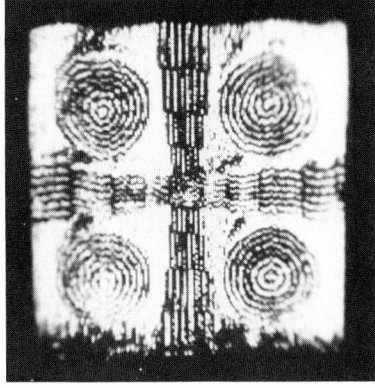

June 1964—Oscar III is being readied for launch with great anticipation because its translator system will allow amateurs to communicate over several thousand miles using only the two-meter band.

Upper left:

1961—Image reproduced from a recording by WA2BCW of a 4030-kc. scfm transmission from A2KPQ. The wavy right edge was caused by wow in the tape recorder. Slow-scan television developed from these beginnings.

Upper right:

LIFTOFF...OSCAR I is on its way. Rapid-fire sequential camera catches launch of Discoverer at Vandenberg Air Force base. Oscar Satellite is in Agena second stage, below black and white checkered band. *(USAF photo)*

Lower left:

During the early morning hours of December 12, 1961 the giant Discoverer XXXVI is prepared for its historic journey. Oscar package was placed in Agena second stage by Bob Herrin (K4RFP/6), Launch Operations Manager. At 2042 GMT the Discoverer was successfully launched into orbit, and shortly thereafter, the 145-MHz "HI" of Oscar I was heard by the hams at KC4USB (Marie Byrd Station) in Antarctica. *(Lockheed Missiles & Space Co. photo)*

Lower right:

Mobiles are lined up and ready to go in the Orange and Louisa counties (Virginia) SET. Left to right are K4DCN, K4LTO, K4CVL, W4SXH and EC K4JYL.

Upper left:

W1CUT's portable station, set up to bounce 1215-Mc. signals from the rocky sides of Talcott Mountain, in the distance. Corner reflector folds for flat carrying. A.c. power is obtained from a 12-volt inverter.

Upper right:

FM communications on 440 MHz.

Lower left:

This is a transistor receiver built by WB6AIG, using an insulated-gate field-effect transistor as a front-end mixer.

Lower right:

K4HDQ doesn't go for these fancy ham shacks... just any l'il tree will do. (photo via K4HDR)

September 1964—The Maxim Memorial Station is undergoing extensive reconstruction, but W1AW is continuing bulletins, code practice and such with makeshift arrangements in the basement.

June 1964—Sixteen societies in the Americas convened (some by proxy) in Mexico City to form the Region II Division of the International Amateur Radio Union, an accomplishment long sought by our League president W6ZH.

★

September 1964—Senator Barry Goldwater, K7UGA, K3UIG, has been named the Republican Party's nominee for the nation's president. We can all dream of seeing a beam atop the White House, but meanwhile, *QST* admonishes us to avoid partisan politics on the air.

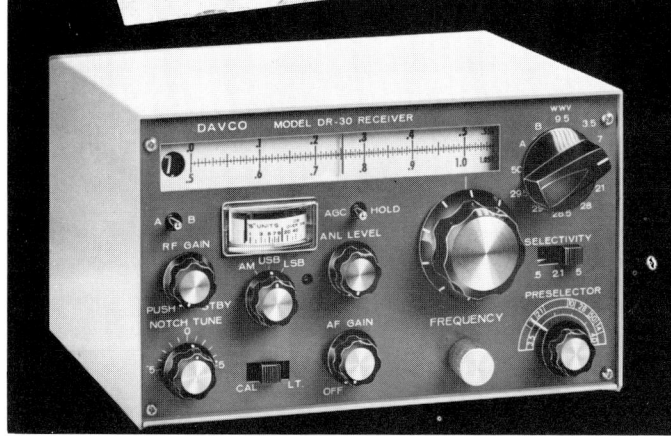

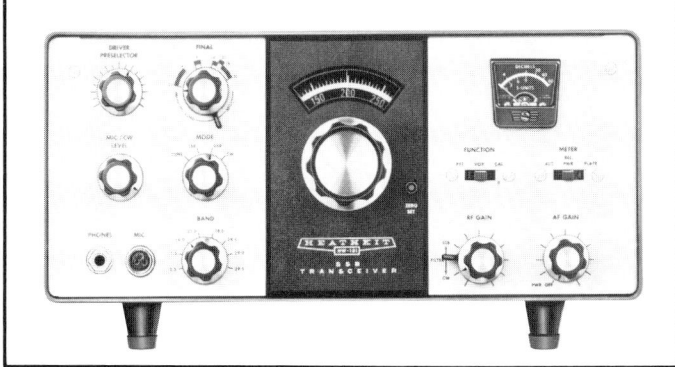

Upper left:

A pair of scissors tripped by pulses bounced off the moon gives Sam Harris, W1FZJ, a beard trim. The signal, a two-and-a-half second pulse, comes from a 1000-watt, 1296-megacycle rig. It was picked up in Dorset, Ohio, after the 523,000-mile trip to the moon and back by W8LIO. The moon pulses were relayed from Ohio to New York City on the 40-meter band to cut the tape for the opening of the 1960 Hudson Division Convention.

Upper right:

W1ISI shows how to make your homebrew project look professional, including how to rivet chassis corners in tight spaces.

Center photo:

The DAVCO DR-30 Receiver, one of the first all-transistor commercial receivers.

Lower left:

Meteors are falling out of the dipper at a rate of 1 per second on this 43 second exposure. This is during the height of the Leonid shower on November 17, 1966.

Lower right:

Heathkit's HW-101 (Hot water 101), a very inexpensive full featured transceiver.

UNCLE DAVE'S RADIO SHACK

FORT ORANGE RADIO DISTR. CO. INC.

USED & NEW SPECIALS

NAT NCX-3 w/NCXA (Used)	$350	JOHNSON INV 2000 (used)	..$875
HAMM HX-500 (used)	350	SB-33 w/SB-LLA LIN (used)	500
NAT NCX-3 (new Demo)	295	HAMM HQ-170A (used)	250
HALLIC SX-101A (used)	295	HALLIC SX-117 (new)	250
GONSET COMM. IV 6 mtrs	350	HALLIC HT-44 (used)	325
COLLINS 32V-3 (used)	295	HALLIC SX-111 (used)	175

904 Broadway, Albany 7, N.Y., U.S.A.
Cable Address: "Uncledave"
CALL Albany 518-436-8411; Nights 518-477-5891

THE BALUN THAT HAS BEEN PROVEN AND ACCEPTED. NOW BEING USED BY THE U.S. NAVY, COAST GUARD, AIR FORCE, ARMY, FCC, CIA, RCA, NBC, FAA AND CANADIAN DEFENSE DEPT. AND BY THOUSANDS OF HAMS IN THE USA AND THROUGHOUT THE WORLD. THE FIRST BALUN WITH A BUILT-IN LIGHTNING ARRESTER THAT COULD SAVE YOUR BALUN AND YOUR VALUABLE EQUIPMENT. The only universal balun with 4 features. No need to buy another type if you should change your antenna in the future. Our one type adapts to all coax fed type antennas. Backed by 50 yrs. of electronic knowhow.

Available at all leading dealers. If not, order direct.

See page 164 for quad ad

W2AU BALUN LETS ENTIRE ANTENNA RADIATE!

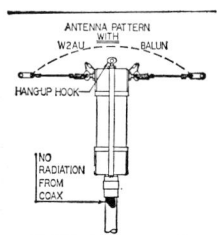

STOP WASTING YOUR SIGNAL! REMEMBER, YOUR ANTENNA IS THE MOST IMPORTANT PIECE OF GEAR YOU OWN.

- No Radiation from Coax
- No Center Insulator Needed
- Perfect for Inverted Vees (Use Hang-up Hook)
- Built-in Lightning arrester
- Broad-Band 2.8 to 40 Mc.
- Takes Legal Power Limit
- Two Models:
  1:1 50 ohm coax to 50 ohm balanced
  4:1 75 ohm coax to 300 ohm balanced
- A must for Inverted Vees, Doublets, Quads, Yagis etc.
- Weighs 6½ oz. 5½" long

(pat. appld.)
$ for $
Your best balun buy.

HELPS TVI PROBLEMS
IMPROVES F/B RATIO
BY REDUCING LINE PICKUP

UNADILLA RADIATION PRODUCTS, Mfrs. of Baluns & Quads, Unadilla, New York 13849

"...IN THE DOG HOUSE?"
MOVE IN
WITH DESIGN INDUSTRIES
WIFE-APPROVED
COMMUNICATIONS DESK

$145.00

... would YOU believe ... SOME hams are permitted into the house ... perhaps even the living room when their station includes a Design Industries Communications Desk or Console?

Send Today For Our Special Wife Pacification Kit
(Descriptive Brochure)

DESIGN INDUSTRIES, INC.

P.O. Box 19406
(214)-528-0150

Dept. T
Dallas, Texas 75219

Congratulations...ARRL

On the Dedication of Your New Headquarters Building. A long record of achievement of which all HAMS may be justly proud.

VIBROPLEX

Also has a long record of achievement, as it is the first of all SEMI-AUTOMATIC BUGS. Vibroplex actually does the work for you. Sends Better, Sends Easier, Sends Faster and Sends Longer, as many keys are still in use after 30 or more years. Most EXPERTS choose the Vibroplex.

VIBROPLEX

America's finest radio key, SEMI-AUTOMATIC and adjustable to any speed. Easier sending than you ever dreamed of. Most experts choose Vibroplex. Standard models have Polished Chromium top parts and gray base; DeLuxe models also include Chromium Base, red switch knob and finger and thumb pieces. Comes in five models from **$17.95** to the Presentation model at **$33.95**; with 24K gold plated base.

VIBRO-KEYER

The finest key made to use with ELECTRONIC TRANSMITTING UNITS. Weighs 2¾ lbs. and has a base 3½" by 4½" with Vibroplex's finely finished parts including 3/16" contacts. With its red knob and finger and thumb pieces it is a thing of beauty. Standard model is priced at **$17.95**; DeLuxe model also includes Chromium Plated Base **$22.45**.

Order today at your dealers or direct

FREE Folder

All Vibroplex keys available for left-hand operation. $2.50 additional. Every Vibroplex has 3/16" contacts.

THE VIBROPLEX CO., INC.
833 Broadway   **New York 3, N. Y.**

W. W. Albright, President

Prices subject to change without notice

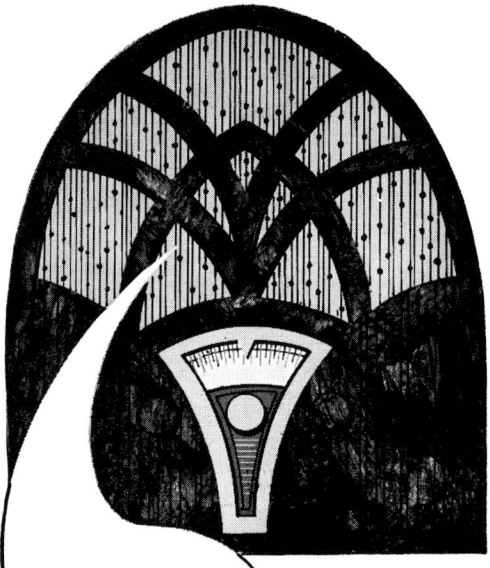

Tedford crystals... since the early days of radio

Back when radio wore short pants, we were one of the gang. Today we're showing our age ... in the capability to produce quartz crystals with remarkable aging characteristics, minimal drift factors and an unusually broad frequency range.

Tedford may be the old timer in the industry, but we don't look it. Our new plant is deceiving — air conditioned, humidity controlled, with the most sophisticated equipment we could lay our hands on.

If you need a few crystals (for your rig) ... or thousands of crystals, crystal filters, integrated filters or temperature compensated crystal oscillators (in your electronic industry job), Tedford is sincerely interested. Write for free product catalogs: Tedford Crystal Labs, Inc., 4916 Gray Road, Cincinnati, Ohio 45232. Phone: 513/542-5555—TWX: 810/461-2476.

TEDFORD CRYSTAL LABS, INC.

FREQUENCY CONTROL DEVICES

THIS QSO WAS SOLID!

One night ... When the skip was perfect on 20 ... I happened to mention the problem of saving money to an OM in 7-land ... how hard it was to save *anything* from the pay check. He said he had licked the problem by banking the *extra money* he earned in *mobile-radio maintenance*.

READING QST LATER, I SAW THE LAMPKIN AD AND SURE AM GLAD I REACHED FOR THE SCISSORS. NOW MY MOBILE-RADIO WORK HAS GROWN TO WHERE ... EVERY MONTH ... I PUT A THREE-FIGURE AMOUNT INTO THE SAVINGS ACCOUNT!

NEW ... THE PPM METER, AN ACCESSORY FOR THE 105-B. ACCURACY 0.0001% FOR SPLIT-CHANNEL FREQUENCY CHECKS. PRICE $147.00, NET.

HERE IS THE SAME COUPON THAT I USED — BETTER MAIL IT NOW!

Lampkin 105-B Frequency Meter. 0.1 to 175 MC and up. Price $260.00, net.

Type 205-A FM Modulation Meter. Range 25 to 500 MC. Price $270.00, net.

```
LAMPKIN LABORATORIES, INC.
MFG Division, Bradenton, Fla.
At no obligation to me, please send me
free booklet "HOW TO MAKE MONEY
IN MOBILE-RADIO MAINTENANCE"—
and data on Lampkin meters.
NAME
ADDRESS
CITY              STATE
```

LAMPKIN LABORATORIES, INC. BRADENTON FLORIDA

VESTO TOWER

Survives 156 mph HURRICANE "DONNA"

Vesto's famous "Hurricane-Proof" Construction is the Reason!

NO GUY WIRES

EASY TO ERECT
Step-by-step instructions given! Can be taken down and moved easily!

HOT DIP GALVANIZED
to last a lifetime!

Prices start at

$149.00

THIS VESTO TOWER WITHSTOOD HURRICANE "DONNA" IN FLORIDA

EASY PAYMENT PLAN! Write for new, FREE Literature!

VESTO CO., INC.
20th and Clay St.
North Kansas City, Mo.

---

2 ANTENNAS IN 1

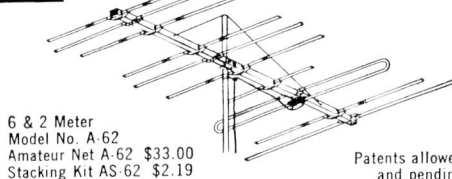

6 & 2 Meter
Model No. A-62
Amateur Net A-62 $33.00
Stacking Kit AS-62 $2.19

Patents allowed and pending

The Only Single Feed Line
6 and 2 METER COMBINATION YAGI ANTENNA

another first from **FINCO®**

ON 2 METERS
18 Elements
1 — Folded Dipole Plus Special Phasing Stub
1 — 3 Element Colinear Reflector
4 — 3 Element Colinear Directors

ON 6 METERS
Full 4 Elements
1 — Folded Dipole
1 — Reflector
2 — Directors

See your FINCO Distributor or write for Catalog 20-226

THE FINNEY COMPANY
Dept. 21   Bedford, Ohio

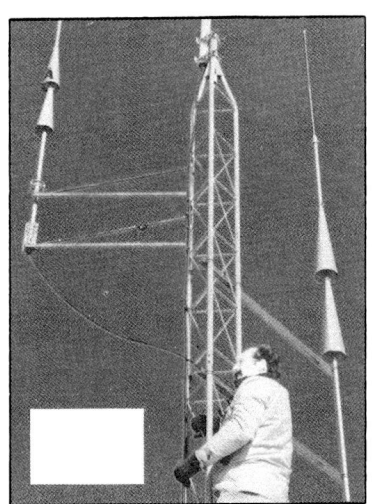

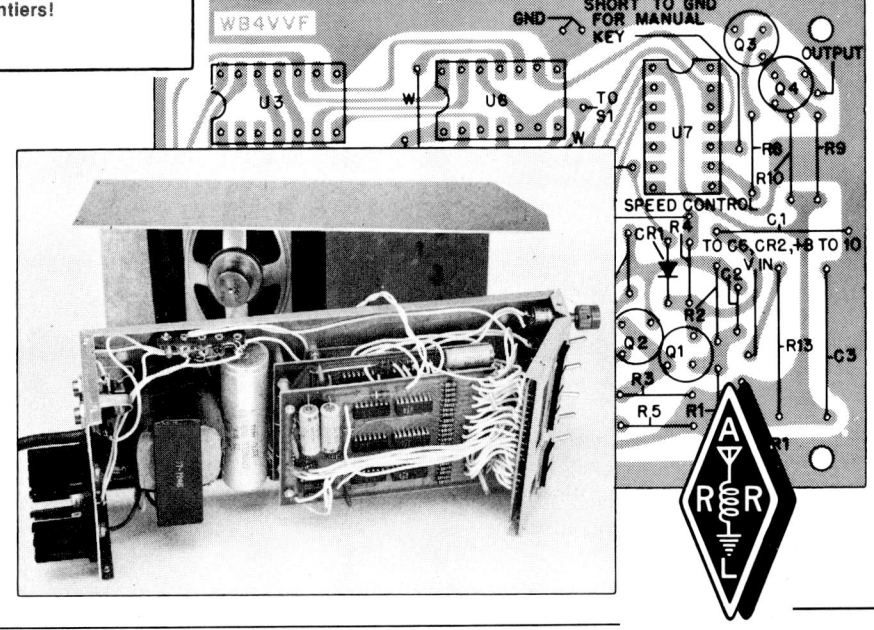

Upper left:

Adella Mueller, N3DQ, demonstrates SSTV to third-grade students in Hickory, Pennsylvania. Each year, Adella and her husband Art, WA3BKD, introduce students to the wonders of Amateur Radio.

Upper right:

WA1MRF "takes out the garbage" with a life-sized high-Q filter. It also functions as an RF oven—doing hotdogs and hamburgers to a turn with 100 watts or more of 2-meter RF power.

Lower left:

Just what a frustrated ham apartment dweller dreams about each February-March when ARRL's annual DX contest comes along! VE7WJ finds enough time left over from maintenance of this all-band antenna farm to serve as president of British Columbia DX Club.

Lower right:

Field Day. Richardson Wireless Klub, K5RWK/5, was a perennial high-scorer throughout the decade.

Upper left:

She's up and adjusted. Now, into the shack and try the new W8JK five-band rotary beam. *(Leo M. Wilhelm photo)*

Upper right:

W1NTH contemplates the differences in size, weight and complexity between a tube-type function generator and a miniature IC generator.

Lower left:

Lolly (W6MAW) and Zeke (W6EOO) Lenn celebrated their 21st anniversary in 1973 with a high jump above Lake Elsinore in Southern California. Which one has the 2-meter hand-held?

Lower right:

This tuna can transceiver is a "purrfect" example of home-brew. Our observer agrees wholeheartedly.

February 1979—Amateur Radio's old archenemy, RFI, is the subject of an FCC Inquiry, General Docket 78-369. What action will the Commission take in solving RFI susceptibility problems of consumer devices? The answer depends largely on whether hams send their comments to the Commission.

Upper left:

An unlicensed Field Day participant—1976.

Upper right, lower left and lower right photos:

WA3UKZ, WB4IMU and 9N1MM show off well-equipped ham shacks of the 70s.

December 1975—Membership in the League continues to grow—up to 120,485 as of the end of October.

December 1976—France now has its first repeater in operation! Its 10-watt signal on European channel "R1" (145.025/145.625 MHz) has reportedly covered all of southern France. Unlike most of European machines, this one is carrier-accessed: no tones needed. Located in the Pyrenees, near Foix, it is a welcome addition to the growing interest in VHF.

December 1977—Her Majesty Queen Elizabeth II has expressed "warm thanks" for the Silver Jubilee greeting sent around the world via AMSAT-OSCAR 7 on June 7. In a letter to AMSAT/UK director Pat Gowen, G3IOR, a representative of Buckingham Palace said the Queen was "most interested to learn of different Jubilee greetings exchanged through the medium of your satellite."

October 1978—The USSR has launched two Amateur Radio satellites named Radio 1 and Radio 2. W1AW OSCAR bulletins carry orbit information.

9N1MM Rev. Marshall D. Moran, S.J.—Kathmandu, Nepal

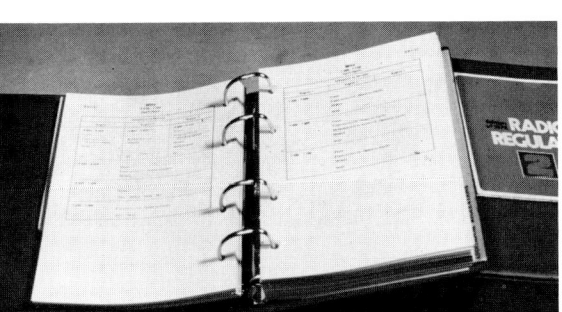

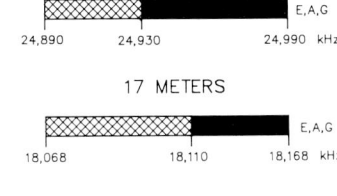

A WARC (1979) at work.

Upper left:

YN1FI was one of more than a score of radio amateurs on their national delegations with specific responsibility for Amateur Radio issues.

Upper right:

Members of the IARU observer team report for duty.

Center photo:

The international Radio Regulations. Revising the Radio Regulations is what a WARC is all about.

Lower left:

ARRL consultant Merle Glunt, W3OKN, carried the ball for US amateurs on our national delegation.

Lower right:

Victory means three new bands!

US AMATEUR BANDS

US AMATEUR POWER LIMITS

At all times, transmitter power should be kept down to that necessary to carry out the desired communications. Power is rated in watts PEP output. Unless otherwise stated, the maximum power output is 1500 W. Power for all license classes is limited to 200 W in the 10,100–10,150 kHz band and in all Novice subbands below 28,100 kHz. Novices and Technicians are restricted to 200 W in the 28,100–28,500 kHz subbands. In addition, Novices are restricted to 25 W in the 222.1–223.91 MHz subband and 5 W in the 1270–1295 MHz subband.

Operators with Technician class licenses and above may operate on all bands above 50 MHz. For more detailed information, see The FCC Rule Book.

KEY
- ▨ = CW and RTTY
- ■ = CW, Voice, SSTV, and FAX

E = AMATEUR EXTRA
A = ADVANCED
G = GENERAL

12 METERS — E,A,G — 24,890 / 24,930 / 24,990 kHz

17 METERS — E,A,G — 18,068 / 18,110 / 18,168 kHz

30 METERS — E,A,G — 10,100 / 10,150 kHz

Maximum power on 30 meters is 200 watts PEP output. Amateurs must avoid interference to the fixed service outside the US.

HUNTER BANDIT 2000C
LINEAR AMPLIFIER

2000 WATTS P.E.P.

- DIRECT-READING WATT METER
- SELF-CONTAINED POWER SUPPLY
- CW/AM/RTTY/SSB
- ALL BANDS—80-40-20-15-10
- GRAY OR BLACK CABINETS

▼

KIT FORM $329.00
(Tubes (8163s) .... $60.00 pair
WIRED AND TESTED $535.00 Complete

Write For Details

Hunter Sales, Inc.
Box 1128E   University Station
Des Moines, Iowa 50311

TYMETER®
"Time At A Glance"

24 HOUR CLOCK

#100-24H

$16

Made in U.S.A.

Walnut or ebony plastic case. 4"H, 7¾"W, 4"D. 110V 60 cy. Guaranteed One Year.

At Your Dealer, or DIRECT FROM

PENNWOOD NUMECHRON CO.
TYMETER ELECTRONICS
7249 FRANKSTOWN AVE.   PITTSBURGH, PA. 15208

Buy your new Hammarlund receiver NOW...

and we'll throw in a free matching speaker.

During December and January, Hammarlund will give you free an extended range speaker, in matching cabinet, with the purchase of any new Hammarlund receiver or linear.

See your Hammarlund dealer for full details.

HXL-ONE

2 K.W., P.P.I., HXL-ONE. Linear amplifier. Compatible with any 100-watt exciter. Grounded-grid: instant-on: Solid state power supply.

HQ-200

BRAND NEW MODEL HQ-200. Versatile general coverage receiver. 540 KHz to 30 MHz in five bands, expanded ham bandspread, SSB product detector, variable B.F.O., Zener diode regulation for superb stability.

HQ-180A

MODEL HQ-180A. Ten to 160 meters in a superlative 17-tube triple conversion general coverage receiver with linear product detector, selectable sideband, and vernier IF passband tuning for unequaled SSB reception.

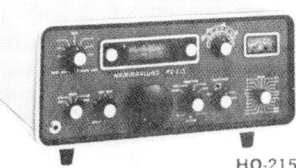
HQ-215

MODEL HQ-215. All solid-state communications receiver. Unequaled sensitivity, selectivity and stability on 10, 15, 20, 40 and 80 meters. Provision for 13 additional 200 KHz segments for general coverage adaptability with communications receiver quality.

The HAMMARLUND Manufacturing Company Incorporated
A subsidiary of Electronic Assistance Corporation
20 Bridge Ave., Red Bank, N.J. 07701

Established 1910

WHY WASTE WATTS?

SWR-1 guards against power loss for $21.95

If you're not pumping out all the power you're paying for, our little SWR-1 combination power meter and SWR bridge will tell you so. You read forward and reflected power simultaneously, up to 1000 watts RF and 1:1 to infinity VSWR at 3.5 to 150 MHz.

Got it all tuned up? Keep it that way with SWR-1. You can leave it right in your antenna circuit.

Swan ELECTRONICS

305 Airport Road Oceanside, CA 92054 (714) 757-7525

DUST COVERS
For Your Equipment

- ADD PROFESSIONAL LOOK
- PROTECT ANY RIG
- MADE OF DURABLE VINYL
- THE PERFECT GIFT
- AS LOW AS $2.95
- CLUB DISCOUNTS

WRITE • COVER CRAFT P.O. BOX 10, ROSELLE PK., N.J. 07204

---

be a big sender.

the hallicrafters co.
A Subsidiary of Northrop Corporation

600 HICKS ROAD
ROLLING MEADOWS, ILLINOIS 60008

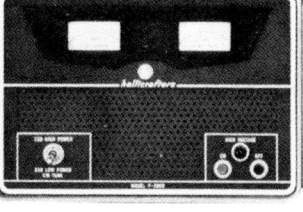  

send big with perfect CW, SSB. lots of watts power. sensational DX'ing. try *hallicrafters* SR-2000 transceiver system. and the power-full line of accessories. put them all together. they make you a big sender. and we mean big. just get a load of this:

HA-1A Keyer
- variable speed 10 to 65 wpm
- digital circuitry for Sidetone
- mercury relay
- transformer operated

SR-2000 Transceiver
- 2000 watts PEP
- less than 1 kHz readout
- built-in RIT, Noise Blanker, AALC
- full metering VOX, MOX, PTT
- P-2000 console speaker power supply

HA-20 Remote VFO
- simultaneous dual-receive
- less than 1 kHz Readout
- built-in VSWR Metering
- incomparable DX'ing Capability
- self-powered

your local hallicrafters distributor has immediate delivery. so go to it, big sender.

PRB – 1

W5LFL and WØORE Operate from Space

Upper left:

Radio-controlled model pioneers and *QST* authors Walt Good, W3NPS/4, and Bill Good, W2CVI, visited League HQ in 1983, prompting this gathering of RC pioneers, current enthusiasts, and staff. The group is standing in the shadow of the famous sailplane, built by R. B. Bourne and Ross Hull, that graces the HQ lobby. Left to right: W2CVI, W3NPS/4 (holding an experimental 926.1-MHz controller), W1CUT, N1BKE, W1DX, WB2PFB, N2BLZ and K1ONG.

Upper right:

K2RIW makes adjustments (100 feet in the air) on his 16-Yagi array for 432 MHz. Dick designed the antennas that are used in this array, and, in fact, his design is used by Amateurs worldwide. Through the use of coaxial relays, Dick is able to limit the number of antennas fed, thereby changing the array beamwidth.

Lower left:

Home-brew equipment used here: frying pan and spatula. Field Day 1982.

Lower right:

"Loyalty" tells the story here. This tower waited until *after* Field Day to topple. *(WA3WIK photo)*

December 1985—ARRL has succeeded in gaining greater protection for 40-meter operators from the transmissions of FCC-licensed broadcasters in the Pacific at 7100 to 7300 kHz.

January 1985—NASA says "go" to WØORE shuttle operation.

Upper left:

K4MF takes to the high seas with a tri-band beam fixed atop this cabin cruiser. *(N4ONQ photo)*

Upper right:

Microwave voice communications on 10 GHz.

Lower left:

Installation of a 145/435 MHz satellite station, which uses a crossed Yagi for VHF downlink and two separate helical antennas for UHF uplink.

Lower right:

In the 1985 ARRL International EME competition, OE9FKI used this 3.8 meter dish on UHF.

December 1985—MARCE in space. The Marshall Amateur Radio Club Experiment (MARCE) will fly on space shuttle mission STS-61C, scheduled for launch on December 18. The experiment will transmit telemetry in synthesized speech directly on 435.003 MHz FM and via AMSAT-OSCAR 10 relay on 145.972 MHz FM.

June 1986—JY2RBH: World's Youngest Ham? Congratulations to King Hussein, JY1, and Queen Noor, JY2NH, on the birth of their fourth child, Her Royal Majesty Princess Raya. In keeping with recent Jordanian tradition, she has been assigned the call JY2RBH.

Upper left:

Operators of the N6CA station made hundreds of 6-meter contacts during the June 1988 ARRL VHF QSO Party. Conditions were phenomenal—note the grid map in foreground. *(N6LL photo)*

Upper right:

Faster than a speeding...well you know! PACKET RADIO

Center photo:

Kathy, WD9DGA, operated in the ARRL Novice Roundup using this Yaesu-equipped station.

Lower left:

This is the largest noncommercial 144-MHz moonbounce array in the world. This station was constructed by W5UN and uses 48 18-element Yagis. The array pivots arounds a center mast, and it uses two automobile chassis as secondary end supports.

Lower right:

W6CPL operates on many VHF-UHF-SHF bands using a combination of home-brew and commercial equipment.

December 1987—The big news from the FCC statistics is the number of new amateurs. In 1987 this number was 24,338, up 40% from 17,848 in 1985, showing that Amateur Radio continues to attract newcomers while inactive licensees don't renew their license.

WØORE—SPACE SHUTTLE CHALLENGER
STS-51 F/Spacelab-2

July 29 to August 6, 1985
First exchange of television images with a manned orbiter
Transceiver: Motorola MX-340, 2.5 watts; *Antenna:* Custom antenna; *Mode:* FM, CW
TV Monitor: Panasonic CT-101; *TV Camera:* Panasonic WV-3050; *Scan Converter:* ROBOT 1200C
Number of young people participating estimated at more than 6,000.
My personal thanks to each of you for making SAREX possible.
73, Tony England, WØORE

W5LFL—SPACE SHUTTLE COLUMBIA
STS-9/Spacelab-1

November 28 to December 8, 1983
First Amateur Radio Station in space W5LFL
Transceiver: modified Motorola MX-300 2-meter FM transceiver, hand-built by the Motorola ARC in Florida 4.5 Watts, FM and CW (by keying carrier).
Antenna: directional ring radiator with cavity, designed to fit in the upper window of the spacecraft; built for NASA by volunteer employees of Lockheed.
Stations: Over 350 two-way contacts, 10,000 SWL cards received, 23 countries.
I am happy you were able to receive my Amateur Radio 2-meter signals from space.
73, Owen K. Garriott, W5LFL

Upper left:

Owen Garriott, W5LFL

Center photo:

Tony England, WØORE

Upper right:

W5DID working on the console supporting SAREX operations. *(NASA photo)*

Lower left:

Family members of Astronaut Tony England, WØORE, during an exchange of imagery from Earth to space using the SAREX hardware and an Amateur Radio station at the Johnson Space Center, August 2, 1985. Kathleen England looks on as her own visage is transmitted to her husband in Earth orbit. Looking on are Amateur Radio operators employed at JSC: Gil Carman, WA5NOM, Lou McFadin, W5DID, and Candy Torres, KA5UKJ. *(NASA photo)*

Lower right:

What do Hollywood producer/director Steven Spielberg and a radio amateur have in common? The answer is the summer 1982 smash-hit motion picture *E. T. The Extra-Terrestrial*. Henry Feinberg, K2SSQ, created the communication device which allowed E. T. to communicate with its home planet. K2SSQ displays the "communicator" he created using a clothes hanger, a child's phonograph, and other household items. E. T. was able to call long-distance—at intergalactic rates, of course. *(Roger Tully photo)*

ARRL ADS THROUGH THE YEARS

NOVEMBER, 1926

The A.R.R.L.
Radio Amateur's Handbook
IS HERE!

FOR five years the production of an A.R.R.L. Handbook has been under consideration. For ten months it has been in preparation. It is now ready for distribution. It is an official publication of the American Radio Relay League. It is endorsed and recommended by that organization as the best and most informative book on short-wave amateur work ever published. Its production has been governed by the same policies of conservative treatment and technical accuracy and clarity of description which have long characterized *QST*. It is the indispensible reference book for every radio amateur or experimenter. You *can't* get the best performance or the most enjoyment from your radio work unless you have the handbook always available. You need the handbook and you will find it the most valuable piece of amateur radio literature ever published.

And it only costs a dollar—

postpaid anywhere

You will never have an opportunity to buy more value for one dollar—send for yours today

AMERICAN RADIO RELAY LEAGUE
1711 PARK STREET   HARTFORD, CONN.

DECEMBER, 1930

Everything that you've wanted in a log is in the Official A. R. R. L. Log Book

New page design to take care of every operating need and fulfill the requirements of the new regulations!

New book form! No more fussing with binders, or trying to weight down loose sheets when the breezes blow!

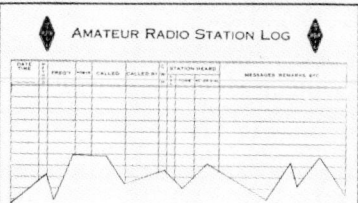

New handy operating hints and log-keeping suggestions, put where they are always convenient!

Designed by F. E. HANDY
A. R. R. L. Communications Manager

THERE are 39 pages like the one above, 8¼" x 10¾", carefully designed to incorporate space for all the essential information you want and need to record about your station's operation. Thirty-nine blank pages (backs of the log pages) to be used for notes, experiments, changes of equipment, etc. Durable covers of heavy stock with space for your station call and dates over which the log entries extend. On the inside covers and first two pages are complete instructions on maintaining your log, convenient tabulations of the most-used Q signals, miscellaneous abbreviations, operating hints, amateur prefixes and signal-strength scales. The information you want, always at your finger-tips.

The new regulations require a log; a well-kept one identifies your station; a uniform series constitutes a progressive and permanent record.

We honestly believe the new Official A.R.R.L. Log Book is the best you've ever seen!

40 cents each   Three for $1.00

Postpaid anywhere

SEND IN YOUR ORDER TODAY!

American Radio Relay League, Hartford, Conn., U. S. A.

GASOLINE and AMATEUR RADIO

A tankful of gasoline in the average car would cost you more than ARRL membership and QST.

———··———

And you can enjoy QST *every* day of *every* month.

———··———

QST and ARRL Membership
$4 in U.S.A., $4.25 in Canada
$5 elsewhere

———··———

**Is Amateur Radio Worth
8 CENTS A WEEK
to You?**

AUGUST, 1951

NOVEMBER, 1962

*P*LANNING new antennas for the Sweepstakes and other contests coming up soon? Looking for dope on transmission lines? From basic theory to how to build 'em, horizontals, verticals, rotaries, fixed beams, transmission lines, together with dimensions, photos, drawings, radiation patterns, you'll find the information in the Antenna Book. Better pick up your copy now.

$2.00 *U.S.A. PROPER*
$2.25 Elsewhere

THE AMERICAN
RADIO RELAY LEAGUE, INC.
WEST HARTFORD 7, CONNECTICUT

TUNE IN THE WORLD WITH HAM RADIO

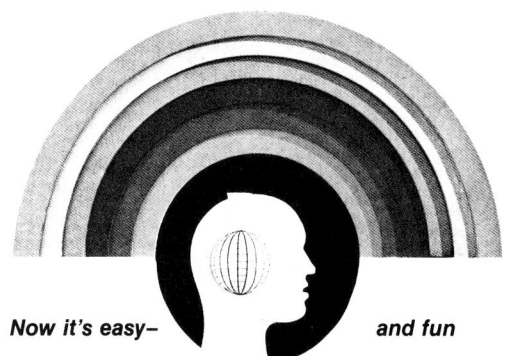

Now it's easy—   and fun

Here at last is everything a beginner needs to know to become a radio amateur—a ham.
It's all in one complete package . . .

- a lively, copiously-illustrated manual
- a Morse code cassette (with an introduction by noted TV star Jean Shepherd)
- an official U.S. call area color wall map.

Everything is covered to fulfill the novice license requirements, assemble an amateur station and get on the air. Best of all it was produced by hams at ARRL, the people who know amateur radio best. **The whole package is just $7.00**—at your favorite electronics dealer or from ARRL headquarters.

Order now from:
THE AMERICAN RADIO RELAY LEAGUE
Newington, CT 06111

MARCH, 1977

FIELD DAY
THROUGH THE YEARS

The Future of Amateur Radio

Contemplation of the future can be a very enjoyable pastime. It can also be most productive. The pioneers of Amateur Radio contemplated the future. If they had not, many of the accomplishments we take for granted today would not exist.

You've just seen the numerous changes and advancements that Amateur Radio has undergone since its beginning. From spark to space in 75 years is an incredible achievement, and a testimony to the creativity of amateurs everywhere.

Before you take your leave, allow us to take you on one more journey...

THE FUTURE: ENDLESS POSSIBILITIES

Packet Radio

Packet radio was the premier amateur technical development of the eighties, at least as measured by its impact on the amateur world. What lies ahead for packet in the nineties? Packet is still very much in its infancy. Faster—much faster—speeds are going to arrive early in the decade. By the mid-nineties, 1 megabit-per-second (Mbit/s) speeds are going to be usual. Many 10-Mbit/s links and networks will be operating, leading to fascinating new possibilities, such as digitized voice and video. A likely scenario, circa 1998: You're driving in New Jersey and want to talk to your friend in Boston. You dial up the local 2-meter repeater and, with a few key strokes on the tone pad, tell the repeater to link to your friend's favorite 2-meter repeater in Boston. Your choice is digitized by your local repeater and sent across the packet network to the repeater in Boston, where it is retransmitted. Receiving no answer (your friend is working country number 345 on 17 meters at the moment), you use a few more keystrokes to tell the remote repeater to store a message. The Boston repeater stores the digitized message on its 2.5-gigabyte disk drive for later relay to your friend. (Or perhaps it sends it to your friend's packet system for local playback!) If your friend is in Seattle instead of Boston, the signal would probably get relayed through a Phase 4 satellite instead of the ground network.

New applications will sprout as the packet system becomes more capable. Less delay

Space Station Artist's concept. Named *Freedom* by President Ronald Reagan, the US Space Station is under development and is scheduled for permanent occupancy in orbit by the mid-1990s. *Freedom* will be an international work place for scientific and commercial research that takes advantage of the gravity-free environment. This artist's concept of the current planned configuration was done for NASA by Alan M. Chichar. Perhaps the final design plans will include an Amateur Radio station. *(NASA photo)*

across the network will make possible applications that are interactive, such as remote data-base servers with full-screen displays on the user's computer. Given the speed with which the capabilities of the personal computer increase, what wonderful things the nineties will bring are literally unpredictable.

In and Beyond the Ionosphere

Amateur satellites have played a big role in the recent history of Amateur Radio. In the nineties, we can expect a quantum leap forward in the communication capability of amateur satellites. AMSAT's proposed Phase 4 satellite will be in a geosynchronous orbit, allowing 24-hour communications without the need for steerable antennas at the ground stations. It will provide both narrowband (voice) and wideband (television and high-speed data) transponders and will use microwave bands, where QRM with other modes won't be a problem. The result will be an amateur communication capability beyond anything previously known. A successful Phase 4 will undoubtedly be followed by other geosynchronous "birds" to provide worldwide coverage.

The nongeosynchronous orbits won't be neglected either. The early-1990 launch of Microsats will be followed by ever more capable satellites in circular and elliptical orbits. The trend is for more people around the world to get involved in building and launching amateur satellites. Some day we may well look back at the nineties as the "golden age" of amateur satellites.

Hams in Space

As the sixties saw the first tentative steps of humankind into space, the eighties saw the first amateurs operating outside the atmosphere. W5LFL and W0ORE operated from the US space shuttle, and Soviet amateurs operated from the MIR space station. Just as the first steps of the sixties were followed by manned exploitation of space, the nineties look to be the decade of expanded manned amateur operations, focusing on *using* the capabilities of a human operating from Earth orbit. Expanded shuttle operations in the early part of the decade

The Elser-Mathes Cup reached ARRL headquarters in 1929, and presently occupies a prominent spot in the museum. The base symbolizes Earth and the seated figures its inhabitants. The bowl is Mars, and the standing men are the amateurs who bridge the gap of space. Mars was chosen as the focal point because the cup's creators, Fred Johnson Elser and Stanley M. Mathes, knew of Hiram Percy Maxim's fascination with the red planet. This unique trophy is still waiting to be claimed.

will exploit the fascination of space-to-classroom communication to interest more young people in both Amateur Radio and the manned space program. Later in the nineties, amateurs hope to place a permanent amateur station aboard the planned US space station. This equipment will include two-way fast-scan television, relayed through the AMSAT Phase 4 satellite, to link the space station with classrooms throughout the US. The station also will be available for crew recreation and experimentation.

Digital Signal Processing

Digital signal processing (DSP) has been around for quite a few years, but amateurs have made only limited use of it. That will change in the nineties. DSP digitizes an analog signal (one which varies continuously in amplitude) and processes it with a high-speed computer chip. The result can be almost whatever the programmer wants. Filters, detectors, mixers, modulators, demodulators—in fact, just about any circuit—can be simulated by the DSP system. Initially, most amateur DSP will be performed primarily on signals at audio frequencies. As the chips get faster, operations at intermediate frequencies will be quite feasible. What does this mean? It means that radios of the nineties will have adjustable CW filtering down to 1-Hz bandwidths—without ringing. You'll be able to notch out an interfering carrier with almost no loss of audio quality on the received signal. Interfering signals that can be characterized simply, such as digital data transmissions, can be eliminated from the receiver audio. Demodulation of digital data and video signals will be done in the receiver—no need for modems—and come out as an RS-232 or video signal. And it's all

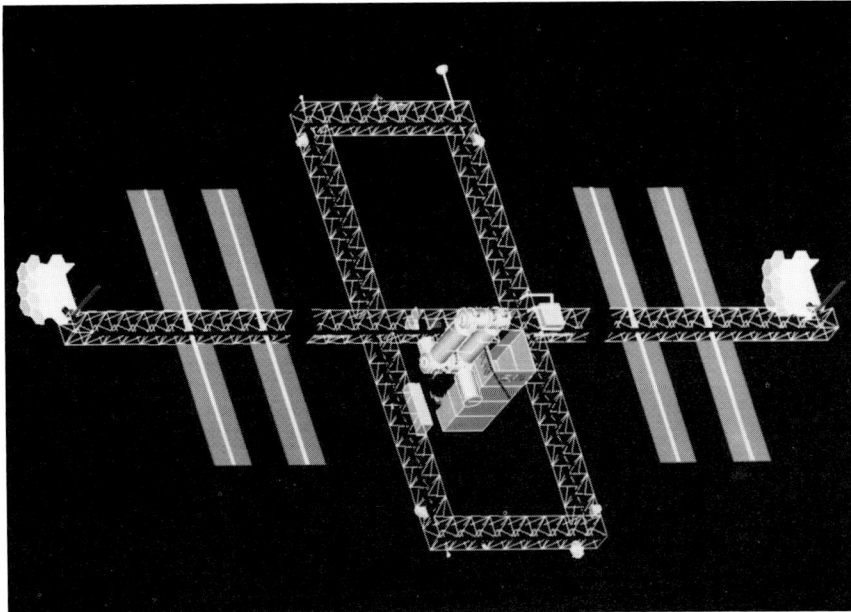

Computer generated drawing of phase 2 of "*Freedom.*" *(NASA photo)*

"only software." How about a "find me a clear frequency" button? Or a "scan for 300-baud packet signals" function? It could happen!

The Microwave Bands

The history of Amateur Radio, and electronic communications in general, has been one of development of ever increasing frequency bands. In the nineties, the pressures of crowded HF and VHF bands, coupled with advances in microwave technology, will make the microwave bands the next frontier to be populated. The line-of-sight nature of microwave signals, together with the extremely high antenna gains that can be realized at microwave, suggests that point-to-point and satellite circuits are the primary applications. Point-to-point circuits will be used to network communications systems, both digital and analog. Wide-bandwidth signals such as high-speed packet and FM television can easily be accommodated at microwave, too.

Obviously, no one can accurately predict the future. We hope you've enjoyed our effort. No doubt, you have a few predictions of your own. There is one prediction we feel fully confident making:

Amateur Radio will continue to thrive, advance, and be an integral part of the future.

Editor's Note: Special thanks to those who lived the history and shared it with us: E. Laird Campbell, W1CUT, Byron Goodman, W1DX, George Hart, W1NJM, John Huntoon, W1RW, and Joseph Moskey, W1JMY.